SpringerBriefs in Energy

SpringerBriefs in Energy presents concise summaries of cutting-edge research and practical applications in all aspects of Energy. Featuring compact volumes of 50 to 125 pages, the series covers a range of content from professional to academic. Typical topics might include:

- A snapshot of a hot or emerging topic
- A contextual literature review
- A timely report of state-of-the art analytical techniques
- An in-depth case study
- A presentation of core concepts that students must understand in order to make independent contributions.

Briefs allow authors to present their ideas and readers to absorb them with minimal time investment.

Briefs will be published as part of Springer's eBook collection, with millions of users worldwide. In addition, Briefs will be available for individual print and electronic purchase. Briefs are characterized by fast, global electronic dissemination, standard publishing contracts, easy-to-use manuscript preparation and formatting guidelines, and expedited production schedules. We aim for publication 8–12 weeks after acceptance.

Both solicited and unsolicited manuscripts are considered for publication in this series. Briefs can also arise from the scale up of a planned chapter. Instead of simply contributing to an edited volume, the author gets an authored book with the space necessary to provide more data, fundamentals and background on the subject, methodology, future outlook, etc.

SpringerBriefs in Energy contains a distinct subseries focusing on Energy Analysis and edited by Charles Hall, State University of New York. Books for this subseries will emphasize quantitative accounting of energy use and availability, including the potential and limitations of new technologies in terms of energy returned on energy invested.

More information about this series at http://www.springer.com/series/8903

Ma Xiaoying · Malcolm Abbott

China's Electricity Industry

Past, Present and Future

Springer

Ma Xiaoying
North China Electric Power University
Beijing, China

Malcolm Abbott
Swinburne University of Technology
Hawthorn, VIC, Australia

ISSN 2191-5520 ISSN 2191-5539 (electronic)
SpringerBriefs in Energy
ISBN 978-3-030-53958-0 ISBN 978-3-030-53959-7 (eBook)
https://doi.org/10.1007/978-3-030-53959-7

This Springer imprint is published by the registered company Springer Nature Switzerland AG
The registered company address is: Gewerbestrasse 11, 6330 Cham, Switzerland

Preface

A lot of modern discussions over energy development around the world involve debates over the impact of carbon emissions and climate change. One often missing element in these discussions and disputes is the role of the world's largest electricity generator and carbon emitter; China. Too little is known about that country's electricity industry in most countries around the world, even though world markets are heavily influenced by the country's substantial demand for coal, oil and natural gas. The purpose of this book, therefore, is to provide some general knowledge of the past, present and future state of China's electricity industry. Both of the authors have a fairly close association with the industry, both in China and abroad, and it is their intention to share as far as possible their knowledge of the Chinese electricity industry. The book is by no means the definitive work on the industry, it being so large and complex that it would be virtually impossible to present it in detail in a single work. Instead it is hoped that readers will be inspired by it to learn more about the industry in China and its impact on world energy markets.

Beijing, China | Ma Xiaoying
Hawthorn, Australia | Malcolm Abbott

Contents

1 Introduction 1
1.1 Introduction 1
References 10

2 The Economics of Electricity 11
2.1 Introduction 11
2.2 The Nature of Electricity 11
2.3 Demand for Electricity 15
2.4 Supply of Electricity 16
2.5 System Balance 17
2.6 Conclusion 19
References 20

3 The Historical Construction of the Electricity Industry in China . . . 21
3.1 Introduction 21
3.2 Background 22
3.3 Socialist Development, 1949–1984 23
3.4 Reform from 1985 25
3.5 Electricity Market Reform and Efficiency 26
3.6 Conclusion 28
References 29

4 Structure and Reform 31
4.1 Introduction 31
4.2 China's Electricity Reform 32
4.3 Natural Gas and Electricity 36
4.4 Conclusion 47
References 48

5 Renewable Energy 51
5.1 Introduction 51
5.2 Renewable Energy Sources 52

5.3 Chinese Electricity Markets Reform and Renewable Energy 57
5.4 Conclusion 64
References 64

6 The Future 67
6.1 Introduction 67
6.2 Background 68
6.3 Energy Demand 71
6.4 Energy Demand Forecasting 72
6.4.1 Simple Descriptive Analysis 73
6.4.2 Factor (or Decomposition) Analysis 73
6.4.3 Advanced or Sophisticated (Econometric) Techniques 74
6.5 Forecasts 75
6.6 Conclusion 79
References 80

7 Conclusion: The Way Forward 83
7.1 Conclusion 83

About the Authors

Ma Xiaoying is a university academic at the North China Electric Power University and has worked extensively in the business sector in China. She holds a Ph.D. in Economics from the Swinburne University of Technology in Melbourne in Australia.

Malcolm Abbott is an economist by profession who specialises in research in energy markets, water supply, transport and network industries in general. In the past he has worked for the ACCC, KPMG and as a Ministerial Advisor. He is currently an Associate Professor of Economics at the Swinburne University of Technology in Australia and holds a Ph.D. from the University of Melbourne.

List of Figures

Fig. 1.1 China's primary energy consumption shares by fuel source 2018 (%). *Source* BP (2019) 3
Fig. 1.2 Electricity generated in China, 1985–2018 (terawatt hours). *Source* BP (2019) 5
Fig. 1.3 Percentage of different ownerships in China's total installed capacity in 2010 (%). *Source* State Electricity Regulatory Commission (2011) 7
Fig. 2.1 World energy consumption by region, 1990–2018 (Million Tonnes Oil Equivalent). *Source* BP (2019) 12
Fig. 2.2 Electricity generation in China, 1990–2040, actual and forecast (thousands TWh). *Source* BP (2019) 12
Fig. 4.1 Natural gas proportion of total energy use in China, 1965–2018 (%). *Source* BP (2019) 35
Fig. 5.1 Renewable energy capacity in China, 1996–2018 (MWs). *Source* BP (2019) 52
Fig. 5.2 Carbon dioxide emissions in China, 1996–2018 (million tonnes). *Source* BP (2019) 53
Fig. 6.1 World energy consumption growth by region; ten-year average, actual and expected (%, per annum). *Source* BP (2019) 68
Fig. 6.2 China's population, GDP and electricity generation growth, actual and expected (%, per annum). *Source* United Nations (2020) Authors' estimates 76
Fig. 6.3 Annual average growth of demand for electricity by sector, 1996–2040, actual and expected (% per annum). *Source* Authors' estimates 77
Fig. 6.4 Proportion of demand for electricity in China by sector, 1996–2040, actual and expected (%). *Source* Authors' estimates 77

Fig. 6.5 Chinese production of electricity 1996–2040, actual and expected (TW hours). *Source* Authors' estimates. 78
Fig. 6.6 Chinese production of electricity, by source 1996–2040, actual and expected (TW hours). *Source* Authors' estimates 79

List of Tables

Table 2.1 Levelised cost of new generation, USA in 2020 ($/MW h) 13
Table 4.1 Population, energy consumption and average energy consumption, various countries, 2018, 2020 (number, millions tonnes oil equivalent, tonnes per person) 37
Table 4.2 Imports of natural gas into China, 2018 (billion cubic metres) 37
Table 4.3 Composition of electricity generation, China and various countries, 2018 (%) 38
Table 4.4 lb of CO_2 emitted per million BTU of energy for various fuels, USA (pounds/million BTU) 38
Table 5.1 Sources of electricity in China, 2018 (TWh, %) 52

List of Pictures

Picture 1.1 Forbidden City in Beijing lit up at night. *Source* Authors' collection 5
Picture 1.2 Provinces of China in 2020 6
Picture 2.1 Chinese transmission lines. *Source* Authors' collection 13
Picture 3.1 Shanghai Power Plant in Yangshupu, 1928. *Source* Published in Hawks Pott (1930). 24
Picture 3.2 Part of the Three Gorges Project. *Source* Authors' collection 27
Picture 4.1 Cooling towers from nuclear power stations in China 43
Picture 5.1 Wind generators in China 54
Picture 5.2 Solar panels and wind turbines in China. *Source* Authors' collection 54

Chapter 1
Introduction

1.1 Introduction

The general purpose of this book is to provide an examination of the state of the electricity industry in China. It does so by looking at its historical development, its current state and its future potential. In doing so, it will also relate this industry to the broader energy sector in China including the use of oil and other energy sources and their development.

Today (2020), the country of China is one of the world's largest energy consumers and producers. As the country has developed over the past forty years, it has experienced a steadily increasing demand for energy, especially for petroleum. In doing so, the country has become very influential in world energy markets. This development in China has been so pronounced that by some measurements, it has become the world's largest consumer of energy and after the USA, the world's second-largest consumer of oil. China was a net exporter of oil until the early 1990s and after, becoming a net importer around that time. Finally, in 2009, it reached the position of the second-largest net importer of crude oil and petroleum products.

Along with growth in demand for oil, other parts of China's energy sector have also developed in recent years. China's natural gas use, for instance, has risen steadily over the past decade, and the country's government has attempted to increase natural gas use and imports through the construction of gas pipelines linking it to neighbouring countries. In addition, liquefied natural gas (LNG) terminals in major ports have been built to facilitate shipped in natural gas. China is also the world's most important producer of coal, as well as the world's largest consumer and importer of this product, accounting for almost one half of global coal consumption. China's rising coal consumption is one of the country's main drivers that has enabled it to become one of the world's largest energy consumers and is linked strongly to the country's swift industrialisation and modernisation. Coal has also fed in strongly to the growth of the country's electricity industry, which became the world's largest in 2011 and is an important factor in China becoming the world's largest emitter of carbon dioxide.

M. Xiaoying and M. Abbott, *China's Electricity Industry*,
SpringerBriefs in Energy, https://doi.org/10.1007/978-3-030-53959-7_1

Underlying all this growth in demand for energy, and electricity, is the central role that China now plays in the world economy. In 2020, China is the world's most populous country (1.4 billion people), and until it was affected by the coronavirus crisis, it had a steadily growing economy. According to the International Monetary Fund (the IMF), the country's annual growth in real gross domestic product (GDP) was 10% per year between 2000 and 2011 and then slowed to 7% per annum for much of the rest of the decade.[1] This slower growth is still relatively strong compared to that of most countries and is expected to continue into the near future, even taking into account the recent international coronavirus crisis. China's political leadership has announced a target GDP growth rate of 7% for the years of the next decade.[2]

In terms of measurement in Purchasing Power parity (PPP) exchange rates, and making use of estimated IMF data, China's GDP surpassed in size that of the USA in 2014. PPP exchange rates make adjustments for the differing costs of goods and services across countries, and so, attempts to show what exchange rates would have to be to buy the same basket of goods in different places at the same time. It should be borne in mind that costs are much higher in the richer countries of the world, and therefore, comparisons of GDP using PPP exchange rates have a tendency to increase the relative size of economies in less developed nations. This means that there is still some contention about whether China's economy is indeed the world's largest, but whatever the conclusions the country today has a very substantial economy, with a very large demand for energy.[3] This economic achievement on the part of China has not come about without there being some difficulties. For instance, the Chinese Government was forced to take action to mitigate the effects of the 2008 global financial crisis, by undertaking a substantial fiscal stimulus package and in 2020 undertook measures to mitigate the impact of the coronavirus on its economy. These packages have been aimed at bolstering the country's investment in infrastructure and demand for industrial goods. In addition, recent trade conflicts with the USA have brought into question in some circles the sustainability of China's high growth rates.

As already mentioned, economic growth in China has slowed in recent years, with growth in production and exports declining and as the Chinese Government has tried to rein in the high debt levels and excessive investment in some markets. The slowing economy has meant that since 2014, the Chinese Government has eased monetary policy—providing medium-term loans to Chinese banks and reduced the banks' reserve requirements and by cutting interest rates. In addition, the Chinese Government has also carried out periodic strategic fiscal stimulus that targeted infrastructure projects.[4]

Related to these, economic developments have been the gradual political changes that have occurred in China in recent years. In China, in March 2013, a new leadership emerged when Xi Jinping became President and Li Keqiang the Premier.

---

[1] International Monetary Fund (2019).

[2] International Monetary Fund (2019).

[3] International Monetary Fund (2019).

[4] Reuters (2015).

Further, at the Third Plenum, that took place in November 2013 the Chinese Government provided an outline of its broad principles of economic reform. At this time, the government decided that it would continue to pursue an incremental policy of economic reform, which was designed to promote a more: 'balanced economic growth', which involved a shift from growth driven primarily by investment in infrastructure, along with the export of merchandise goods, towards one with a greater level of domestic consumption, including that of services.

In the case of the energy sector, the Chinese Government has tried to shift towards the greater use of market-based pricing, as well as introduced measures that encourage investment in technologies that embody greater energy efficiency as more effective control of pollution. Furthermore, the Chinese Government has also tried to encourage competition between energy companies, as well as made more substantial investments in more challenging sources of electricity such as hydroelectricity and renewable energy. The Chinese Government has in addition also tried to find ways to attract more private investment into the energy sector by carrying out a streamlining of approval processes and by introducing policies that enable more capital expenditure on electricity and gas transmission lines, pipelines and terminals. Despite these changes, however, much of the country's energy sector is still dominated by a small number of very large, government-owned institutions, which has meant that there is still considerable pressure to reform the sector further.

In order to gain more perspective of these changes, it is possible to look more specifically at the mix of energy sources in China. In doing so, we find coal provided for the main part of energy consumption (59%) in 2018 (see Fig. 1.1). After coal, the second-largest energy source is petroleum and other liquids, which accounts for

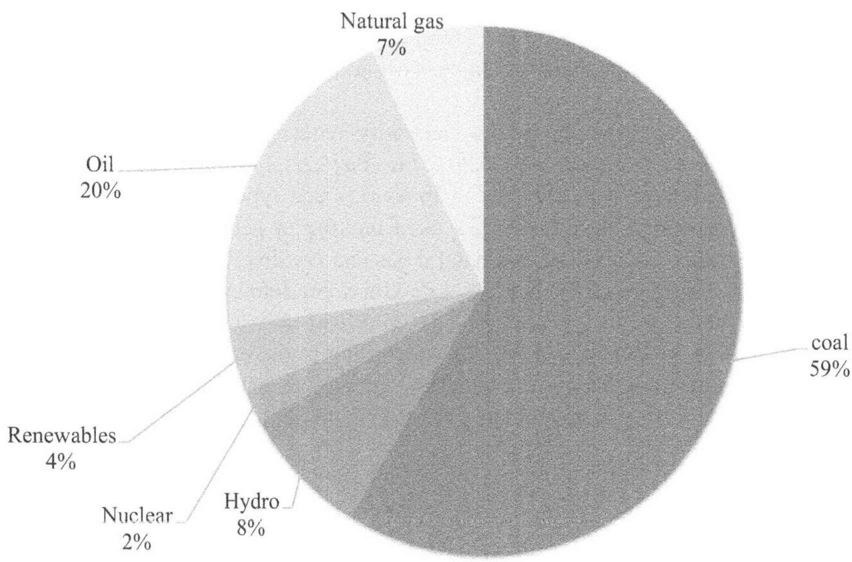

Fig. 1.1 China's primary energy consumption shares by fuel source 2018 (%). *Source* BP (2019)

20% of the country's total energy consumption. Even though the country has tried to diversify its energy supplies sources, those like hydro-electricity makes up only 8% of the total, natural gas only 7%, nuclear power nearly 2% and other non-hydro-renewables more than 4%. Non-fossil fuels, therefore, contribute still relatively small shares of China's energy consumption.

In response to this state of affairs, the Chinese Government has put into place plans to cap coal use at only 62% of the total primary energy consumption in 2020 in order to reduce the heavy air pollution that has afflicted many regions of the country, but also to reduce the country's dependence on imported coal and oil.[5] The Chinese Government's National Energy Agency has claimed that coal use has fallen in recent times, but not by much.[6] In order to cap this coal use, the Chinese Government set a target of raising non-fossil energy consumption (including hydro and nuclear) to 15% in 2020 and to 20% by 2030.[7] In addition to this process, China is currently increasing its use of natural gas to replace some coal and oil use as the former is a cleaner-burning fuel compared to the latter two. In preparing its plans, it was envisaged that the use of natural gas would account for 10% of its energy consumption in 2020, a target that has been achieved.[8]

Even though the absolute level of coal consumption is expected to rise over the longer run as the economy continues to grow and overall energy consumption rises, higher levels of energy efficiency as well as expansion of the use of natural gas, hydro, nuclear and renewables are likely to lead to a decrease in coal's overall share of energy use. Some of these measures will help to mitigate China's position as the world's leading energy-related CO_2 emitter. The Chinese Government has made a commitment to reducing carbon emissions mainly in the energy-intensive industries and in construction. Despite these intentions, the Chinese Government has projected that its carbon emissions will rise by more than one-third from levels currently experienced and will not peak until 2030.[9] This does assume that intentions of the Chinese Government are realised and the country's economy can reduce its dependence on coal.

Turning more specifically to the electricity industry, the Chinese electricity industry became the world's largest in 2011 (see Picture 1.1). In that, country coal and hydro-electricity have long been the main sources and types of installed capacity. In more recent years, the country has been moving to generate more electricity from other sources such as nuclear, natural gas and non-hydro-renewables partly to help expand overall capacity, but also to address environmental concerns already mentioned. Net electricity generation in China was an estimated 5000 Terawatt hours (TWh) in the 2010s (see Fig. 1.2) according to the estimates of BP and the Energy Industry Administration of the USA. This has meant that electricity generation in China has more than doubled since 2005, although the growth in electricity

[5]Du and Jiang (2015) and Dupuy and Wang (2016).

[6]NEA (2015).

[7]NEA (2015).

[8]NEA (2015).

[9]United States, Department of Energy, Energy Information Agency (2020a, 2020b).

Picture 1.1 Forbidden City in Beijing lit up at night. *Source* Authors' collection

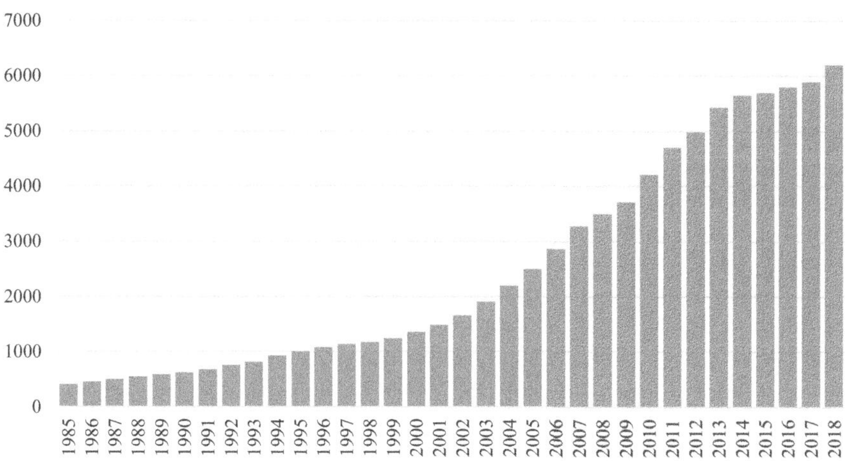

Fig. 1.2 Electricity generated in China, 1985–2018 (terawatt hours). *Source* BP (2019)

generation, which is mostly driven by economic and industrial demand, tended to deaccelerate after the global financial recession in 2008 and 2009.

The importance of the industrial sector to demand for electricity can be illustrated by the fact that it currently accounts for almost three-quarters of the country's

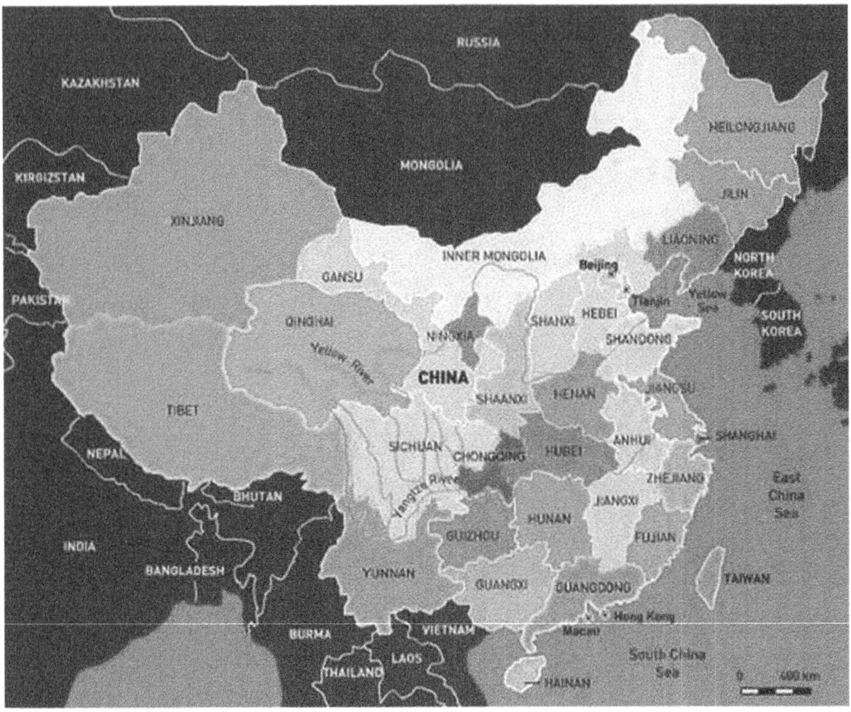

Picture 1.2 Provinces of China in 2020

electricity consumption.[10] The growth in electricity generation per annum was a decade-low 4% in 2014 and has continued since then at close to this level (for annual growth in Chinese electricity generation see Fig. 1.2). This relative decline in the growth of electricity demand was a consequence mainly of the slowdown in growth of activity in the heavy industrial sector of the Chinese economy, in particular that of the steel industry. As previously mentioned, the Chinese Government plans in the future to depend more on electricity generated from natural gas, nuclear and non-hydro-renewable sources. In doing so, the government hopes it will be able to replace some coal-fired plant, with the aim of reducing the country's high level of carbon emissions and urban air pollution.

Installed electricity generation capacity in China in 2020 is an estimated 2010 gigawatts (GW). Installed capacity is expected to be increased in the next few decades to meet rising demand for electricity, especially demand that comes from large urban areas in the southern and eastern regions of China (see Picture 1.2). The Energy Information Administration of the US Government (the EIA) has projected that installed generation capacity will increase in size to around 2265 GW by the year 2040, much of which will be met by the creation of capacity from coal, natural gas, nuclear and renewable sources.

[10]United States, Department of Energy, Energy Information Agency (2020a).

Much of this new capacity will be created by Chinese Government-owned companies, although efforts have been made in recent years to encourage new private initiatives. The electricity industry in China today, despite attempts to attract private investment, is mainly controlled by state-owned holding companies. Reform of the Chinese system has been undertaken with the aim of improving system efficiency and of facilitating investment in the power grids, although the reform has tended to take place at a relatively slow pace.

The present-day structure of the electricity industry was largely formed in 2002 when the Chinese Government disaggregated the monopoly State Power Corporation into separate generation, transmission and services units. Since this reform, generation in the Chinese electricity industry has been controlled by five state-owned generation firms: the China Huaneng Group, China Datang Corporation, China Huadian Corporation, China Guodian Corporation and China Power Investment Corporation. Together, these five firms generate nearly one half of electricity in the country. Most of the remaining electricity is generated by local government-owned enterprises or by independent power producers (IPPs), often in partnership with privately listed subsidiaries of the state-owned firms. Partial deregulation of the industry and other reforms has opened up the industry to some investment from overseas, although this investment has tended to date to be fairly limited (see Fig. 1.3).

At the time of the reforms in 2002, the State Power Corporation concentrated its electricity distribution and transmission networks into two new firms, the State Grid Corporation of China and the China Southern Power Grid Company. These two companies between them operate the country's seven power grids. The State Grid Corporation manages transmission grids in the north and central regions of

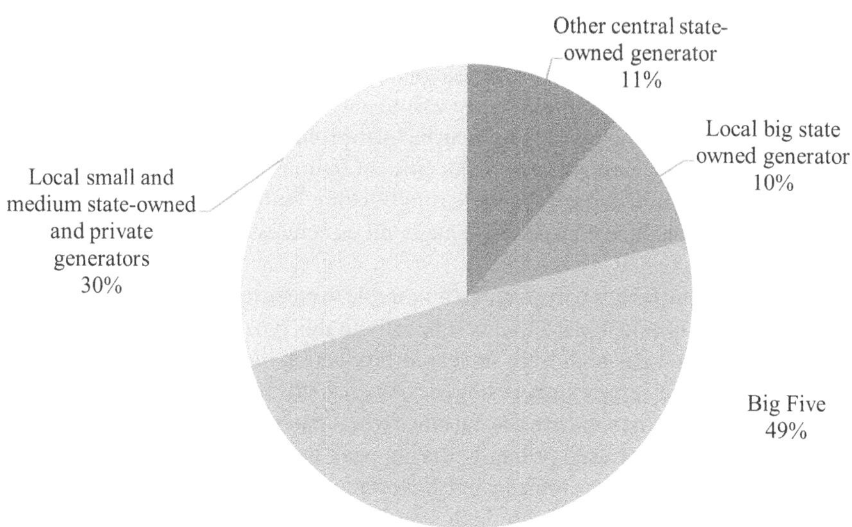

Fig. 1.3 Percentage of different ownerships in China's total installed capacity in 2010 (%). *Source* State Electricity Regulatory Commission (2011)

the country, while the China Southern Power Grid Company manages those in the southern part of the country. At the same time, the Chinese Government also established the State Electricity Regulatory Commission and made it responsible for regulation of the industry and for the facilitation of new investment in the industry. In later years, as part of the Chinese Government's efforts to streamline government agencies, in March 2013, it abolished the State Electricity Regulatory Commission and transferred its responsibilities to the National Energy Administration (NEA).

The NEA in turn is an agency of the National Development and Reform Commission (NDRC). In China, electricity sold by generators to the grid, as well as retail electricity prices, is determined and capped by the NDRC, which also sets the price that coal companies should receive from generator companies. The Chinese Government began to reform electricity prices in 2004, and after the 2002 restructuring, by introducing a policy that passed through fuel costs, however, on-grid prices were only infrequently altered. As a result of the mismatch, in 2011, high prices of coal and low government-controlled electricity tariffs meant that the generator companies made substantial financial losses. In 2012, coal prices declined and have since remained relatively low alleviating this problem somewhat. Similar reforms in the natural gas pricing mechanism also took place, with electricity tariffs for gas-fired plants being linked to higher natural gas prices. In addition, the NDRC doubled the surcharge on renewable energy use to end-users (except residential and agriculture consumers). This measure was designed to promote investment in renewable energy infrastructure and to facilitate a shift towards using alternative fuels.

The shift towards alternative fuels is important as the Chinese electricity industry is still dominated by coal as a source of fuel. The large amount of domestic reserves of coal in China means that it will remain the most important fuel source for many years to come. To improve the efficiency of coal-fired plant, the Chinese Government is closing small and inefficient plants and modernising the remaining units. Larger, more efficient units as well as technologically advanced ultra-supercritical units, which operate at the highest levels of pressure and temperature for a coal plant, are being favoured. The Chinese Government has also prohibited firms from building new coal-fired power plant around three major cities of Beijing, Shanghai and Guangzhou. In addition, coal-fired capacity is increasingly being built close to the inland coal-producing areas and then transported to major urban centres over expanded electricity transmission lines.

To date, natural gas has only played a minor role in electricity generation; however, the Chinese Government plans to invest heavily in this type of plant. The country is currently obtaining gas both from increased production of domestic sources of gas and from alterative several import sources. As coal still remains the less expensive fuel (except in the large southern coastal cities where natural gas is more competitive), natural gas tends to be used primarily during peak demand for power. There are a number of large natural gas projects in China some being undertaken in conjunction with the opening of new LNG terminals such as those in Guangdong and Shanghai. In Beijing, as well, authorities are replacing coal-fired facilities, with lower emission gas-fired plants (having closed three of the four major coal-fired plants by 2015). In the long term, whether China will be able to shift to gas-fired generation will depend on whether it can produce more from domestic sources of gas, to bring in more imports and to construct a more substantial gas transmission grid. For these reasons, other sources are also being developed.

Nuclear generation contributes a relatively small proportion of the country's electricity. The Chinese Government, however, is actively promoting nuclear power as a lower emission creating source, as well as being relatively low cost. Nuclear capacity has in recent years been greatly expanded. China's nuclear plants are located along the east coast and in the southern regions of the country, but the government plans to assess the construction of inland facilities, according to its latest energy strategy plan. After the Fukushima Daiichi nuclear accident in Japan in March 2011, the Chinese Government suspended approvals for new nuclear plants until safety reviews of all facilities were completed and a safety framework was approved by the State Council. In October 2012, new plant approvals were introduced, and since then, the commissioning of new plant has risen. The Chinese Government plans to boost operational nuclear capacity to 58 GW and has 30 GW of capacity under construction in 2020. This includes the encouragement of more private investment in nuclear project development and the greater use of more speedy approval processes.

Besides natural gas and nuclear energy, the Chinese Government has a stated goal of producing at least 15% of its overall energy consumption from non-hydro-renewables. This has meant that China is now the world's largest investor in renewable energy. The key source of renewable energy in China is hydro-electricity because of its cost-effectiveness and sizeable resource potential. China is the world's largest producer of hydro-electricity, with installed generation accounting for more than one-fifth of total installed generation capacity. The world's largest hydro-electricity project, the Three Gorges Dam along the Yangtze River, was completed in July 2012 and includes 32 generators with a total maximum capacity of 22.5 GW. Another substantial hydro-electricity dam, and China's third-largest - Xiangjiaba, entered operations in 2013. The Chinese Government plans to increase hydro-electricity capacity to 350 GW by the end of 2020.

In terms of non-hydro-renewables, China is one of the world's largest wind producers of electricity, a source that has grown in importance since 2005. The Chinese Government has encouraged investment in grid development and measures to improve flexibility in the transmission system, especially during peak hours in order to enable greater use of non-hydro-renewable electricity. The country is also investing in solar power and hopes to increase capacity from 15 GW at the end of 2013–100 GW by the end of 2020. Besides, wind and solar biomass use in China is relatively small and is mostly used for cooking and heating in rural areas, and for small-scale power projects. The NDRC has created price and tax incentives for investments in biomass and waste incineration projects through feed-in tariffs.

All of this activity is taking place at a time of considerable interest in the nature of energy and its supply, not just in China but in many countries around the world. Increasing our understanding, therefore, of the development of China's energy sector is important to understand how that country is evolving over time. In subsequent chapters of this book, a range of topics will be covered. In the next chapter, a description is given of the historical development of the Chinese electricity industry. The chapter that follows provides more detail on the fuel mix used in China as well as some background on the economics of the electricity industry. Chapter 4 looks at the structure and reform of the industry in China, and this is followed by a chapter on the developments in the renewable energy industry in that country. Chapter 6 provides some information on what might be expected in the future in the industry, and in the final chapter, some conclusions are made.

References

BP. (2019). *BP statistical review of worked energy*. London: BP.

Du, Y. F., & Jiang, Q. (2015). *NEA officials answers reporters' questions about how the electricity reform will push forward integration of renewable energy* (in Chinese). http://suo.im/1MkOsR.

Dupuy, M., & Wang. X. (2016, April 8). *China's string of new policies addressing renewable energy curtailment: An update* (in Chinese). http://suo.im/2eJawp.

International Monetary Fund. (2019). *IMF staff country reports: China*. Washington DC: IMF.

NEA .(2015). *Officials from NEA answered correspondents' questions on electric power sector reforms* (in Chinese). Beijing: NEA. http://suo.im/MWZx.

Reuters .(2015, 5 May). https://www.reuters.com/article/us-china-economy-policy/china-looks-to-fiscal-stimulus-to-fight-slowdown-sources-idUSKBN0NR2FW20150506.

State Electricity Regulatory Commission. (2011). *Electricity regulatory annual report 2010*. Beijing: SERC.

United States, Department of Energy, Energy Information Agency. (2020a). *Annual energy outlook 2019*. Washington: EIA.

United States, Department of Energy, Energy Information Agency. (2020b). *Country analysis briefs: China*. Washington: EIA, May.

Chapter 2
The Economics of Electricity

2.1 Introduction

The electricity industry in many countries since the 1990s has been the subject of significant change. In the process of this change, the industry's operations and decision-making in most of these countries have been altered from being one that is state-dominated and centrally planned, to one where private-oriented decisions have become more important. In many cases, these changes have made decision-making more complex, and this additional complexity has meant that understanding the nature of electricity markets is in some ways becoming more difficult.

It is, therefore, important that a basic understanding of the nature of the industry is acquired before these changes can be studied. This chapter, therefore, will provide a general coverage of the traditional style of electricity system operation and decision-making that exists in the industry. In doing so, indications will be provided where restructuring of the industry can occur and how this has affected decision-making. In subsequent chapters, it will be possible to see how the changes in the Chinese electricity industry have been affected by the basic character of electricity generation and distribution.

2.2 The Nature of Electricity

Energy consumption has grown at a steady rate around the world (see Fig. 2.1), and in more recent times, the importance of the Asia-Pacific countries has become more pronounced. Most estimates of future demand growth for China anticipate continued growth of demand in the future, although this growth will vary across the use of different types of fuels (see Fig. 2.2 for estimates by the company BP). In the Chinese case, the use of various fuels for the production of electricity and energy more generally has grown substantially. More specifically, the mix of fuels in the generation of electricity has changed over time, and this change is expected

M. Xiaoying and M. Abbott, *China's Electricity Industry*,
SpringerBriefs in Energy, https://doi.org/10.1007/978-3-030-53959-7_2

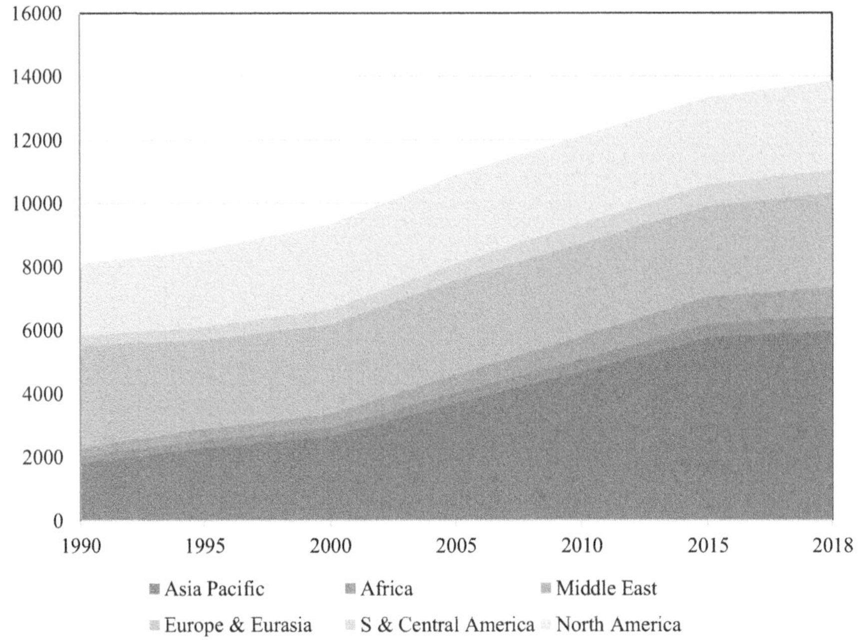

Fig. 2.1 World energy consumption by region, 1990–2018 (Million Tonnes Oil Equivalent). *Source* BP (2019)

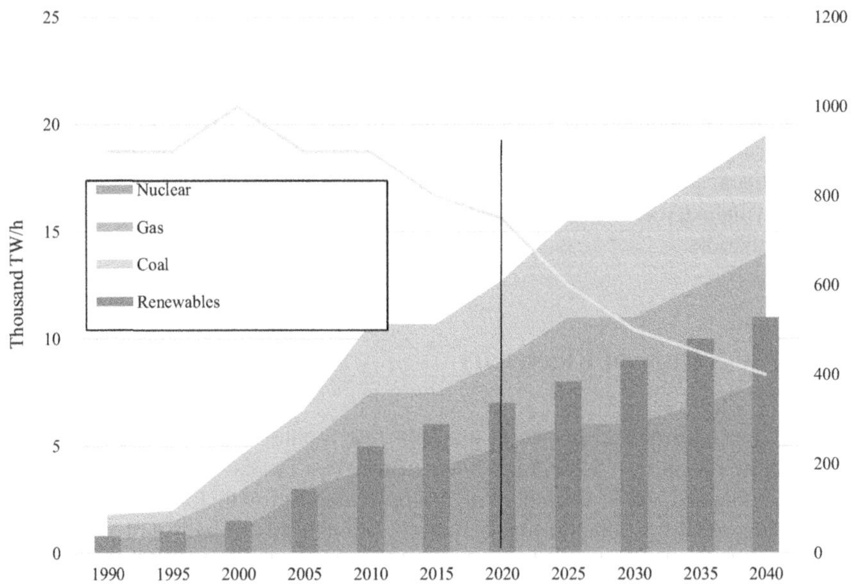

Fig. 2.2 Electricity generation in China, 1990–2040, actual and forecast (thousands TWh). *Source* BP (2019)

to continue into the future. This mix of fuels is influenced by a range of factors, including the relative costs of different types of generation (see for instance the figures in Table 2.1), and how these relative costs change over time.

Table 2.1 Levelised cost of new generation, USA in 2020 ($/MW h)

	Capital cost	Fixed O and M	Variable O and M	Transmission cost	Total system cost
Ultra-supercritical coal	47.57	5.43	22.27	1.17	76.44
Combined cycle	8.4	1.59	26.88	1.2	38.07
Advanced nuclear	56.12	15.36	9.06	1.1	81.65
Geothermal	20.38	14.48	1.16	1.45	37.47
Biomass	39.92	17.22	36.44	1.25	94.83
Wind, onshore	29.63	7.52	0	2.8	39.95
Wind, offshore	90.95	28.65	0	2.65	122.25
Solar photovoltaic	26.14	6	0	3.59	35.74
Hydro-electric	37.28	10.57	3.07	1.87	52.79

Source United States, Department of Energy, Energy Information Agency (2020), *Country Analysis Briefs: China*, Washington: EIA, May, www.eia.doe.gov. United States, Department of Energy, Energy Information Agency (2020), *Annual Energy Outlook*, Washington: EIA

Picture 2.1 Chinese transmission lines. *Source* Authors' collection

In looking at the economics of the electricity industry, this chapter focuses on the grid-based supply of electricity, which is the main form of electricity supply, although in many places, the off-grid supply of electricity is gaining in importance, particularly in remote regions of developing countries. To begin with, it is worth noting that electricity is a form of energy that is difficult to store in any economically viable way. It is instead used by consumers at the time it is produced. This means that when an electricity user is connected to an electricity network and then switches on an appliance, then the demand for electricity is felt in the grid. This demand from individual consumers is aggregated for an electricity system together to make up all system demand. Although consumers can use different appliances that use electricity the tendency is for them not to use all of the simultaneously. Similarly, all consumers together that are connected to an electricity grid do not create all of their demand for electricity at the same time (Picture 2.1).

The location and design of network assets (the distribution and transmission wires that transport electricity) have been defined largely by where the production sources (e.g. coal mines, dams, gas reserves) and large user groups are situated (generally metropolitan centres, sometimes regional areas where large commercial consumers are present, e.g. steel and aluminium smelters). As such, the geographical layout of transmission systems tends to consist of lengthy point-to-point assets, although as systems are developed the inter-linkages between various regions has occurred in most countries. This has meant a greater emphasis which gets placed on developing networks with multiple connections to improve resilience and system stability.

The effectiveness with which a wholesale energy market may function is dependent on the interrelationship between how it is designed and the characteristics of the generation plant serving that market. Coal-fired electricity generators, for example, are generally large-scale infrastructure assets that operate most efficiently when used at a consistent rate. Similarly, closed-cycle natural gas plants are designed for extended periods of use. Such assets may be adversely impacted by a market that requires them to be powered down or even shut off at short notice. By contrast, open-cycle gas plant and dam-fed hydro-electric facilities tend to be more suitable for the provision of intermittent power supply, and a competitive wholesale market is more likely to be made available when both demand and anticipated prices are high. In all instances, however, the timing of use of such generation plant referred to above is controllable. This compares to plant reliant on renewable energy sources such as solar and wind, whose marginal cost of production is low to negligible, yet whose capacity to generate depends on non-controllable events such as the weather and, in the case of solar, the time of day. The mix of generation plant can impact on the efficiency with which a wholesale electricity market operates, while conversely, the design and outcomes of such a market will create incentives for how investment in new electricity plant will be directed.

After the electricity has been generated, it is sent through high-voltage transmission lines to the regions in which it is to be consumed. Energy is transmitted at higher voltages to reduce losses that would otherwise occur. This is because higher voltage is accompanied by lower current for any given amount of power produced, which means resistance is reduced as electricity flows across the network (Ohm's law).

Lower resistance also means lighter cables can be used in long-distance transmission, reducing the cost of both the cables and the towers supporting them. The majority of transmission networks operate utilising alternating current, although some inter-state connectors are direct current (also some underwater links). Once transmitted, the electricity is then transformed to a lower voltage and sent through a distribution network to final customers.

The design of voltage transformations from transmission to distribution (and then to end usage) is a function of the efficient transfer of energy over distance relative to the number and costs of transformers that would be required across a network to supply electricity at a voltage which is practical and efficiency for household, business and commercial users. Larger, more expensive transformers are required to transform the voltage. For the majority of users, the electricity will then need to be transformed again to a lower voltage at the premises, at which point it will be suitable for use. While some small divergence around these levels is permissible, if the voltage or frequency varies too widely, the system will become unstable and there is risk of damage to equipment and appliances connected to the system.

2.3 Demand for Electricity

As well as the simultaneous nature of electricity creation and consumption, it is important to understand the diversity of demand. In a formal sense, the diversity factor can be defined as being the ratio of the sum of maximum customer demands in a system to the maximum system load. The more diversified the load is, the lower the peak capacity requirement is. If demand is diverse, then there is a reduced investment in generation capacity. The reverse of the system diversity factor is known as the coincidence factor.

Generally, it is apparent that the pattern of electricity load on weekdays is different from weekend load patterns. During weekdays, there is greater demand that flows from commercial and industrial activity. As well the demand in the system differs according to the seasons of the year. Generally, in relatively cold countries, demand increases during winter, while in dry, hot or tropical countries, demand for electricity is more pronounced in the summer days of the year. Information about the daily load curves can be collected over the course of a year, in which case the frequency of occurrence of different loads can be estimated. The creation of a plot of such a cumulative frequency distribution by load is known as a load duration curve. For 100% of the time, the system load is more than that of a small amount of load, which is known as the base load. The base load is found on the right-hand extreme of the curve. For 0% of the time, the system load exceeds the peak load (the highest load).

A load duration curve typically has three segments of it. Generally, the system faces a relatively small load. The system has the highest level of demand for only a shorter period, which is known as the peak period. Between the peak and base load periods, the demand gradually increases and decreases between these two loads. This period of increasing demand is known as the period of intermediate load.

In looking at a load duration curve, there is a significance for the operation of plant, the cost of service and the efficiency of the system. It was stated earlier that electricity cannot be easily stored in significant quantities at low cost; this means that demand for electricity has to be met by modulating the supply to match the demand.

This means that for the smooth operation of the system, three types of generation plant (or technologies) are needed. These are as follows.

- Generation plant that runs all year—all the time. This is to meet the base demand. This plant normally does not have the capability to vary the supply depending on demand.
- Other generation plant is needed to follow demand and vary output frequently.
- In addition, a third set of plant is needed which is suitable for running only during peak periods.

Base load generation plant requires technologies with low operating costs, but also might have high capital costs. Generation plant with high operating costs and low capital costs, on the other hand, is generally more suitable for use as peaking plant, as they run for only short periods of time. Intermediate load generation plant often incurs additional wear and tear and loses some efficiency in following the load.

These three different types of plant have different capacity utilisation rates, which are called capacity factors. The base load generation plant is often used nearly 100% of the time. In contrast, peaking plant is used only for very short periods of time (usually 20% of time). If the load did not vary so much during the day, the week, or the time of the year, then generation plant would not be so varied in type.

In light of the change in demand for electricity over time, the shape and size of the three elements of the load duration curve and the overall capacity utilisation are determined. This in turn is known as the system load factor and is the ratio of area under the load duration curve compared to the area of the rectangle formed by the peak load for entire duration of the year.

2.4 Supply of Electricity

Electricity can be made by making use of a variety of different technologies that use alternative fuels. These fuels can be categorised into two basic types.

- Conventional: these types can be further grouped into two categories: hydro and thermal. Hydro-electricity makes use of energy stored in water (i.e. the potential energy) and uses it to generate electricity. Thermal types make use of the chemical properties of the fuel either by burning the fuel and using the heat to create steam from water and then passing it through turbines or in gas combined cycle plant where both gas turbines and steam turbines are used. In the case of nuclear power, the chemical properties of the fuel are used to heat water, which is then used to generate electricity.

- Non-conventional: these technologies include geothermal, solar, wind and other similar types. A few of these types of plant can operate continuously while others can only operate intermittently. Those technologies that make use a stock of fuel normally can operate continuously, while those that make use of a flow only as long as it exists (e.g. tidal, solar and wind) are intermittent. The technologies that make use of the intermittent energy sources must be used when they are generating electricity, taking preference over the other forms.

Each technology has its own constraints in terms of its operating characteristics. For instance, a hydro-storage facility can be brought online relatively quickly and only takes a short period before it reaches peak levels of output (a quick response plant), which means that it can be used as a peaking plant. In contrast, plant that uses nuclear or coal takes a relatively longer time to start up and then shut down and so therefore is not used to supply the fluctuations in load or system frequencies. Natural gas plant can be used relatively flexibly as it can be brought online fairly quickly and generally has a high level of fuel efficiency. When plant has to be shut down because of low demand or operational problems (outages) outages, they normally cannot be brought back online without some delay. In doing so, the generation operator will incur some additional costs each time.

The extra capacity required to be maintained in addition to the level of demand at any point in time is known as the generation capacity reserve. This reserve enables the system to backup any generation plant outages and errors that have been made in demand forecasting and any other errors of faults that occur. The reserve comes in two types.

- Quick-start reserves: this is the plant that can be started up swiftly to create load to meet any change in demand. Natural gas turbines, for instance, can generally be used within a few minutes or starting up. Hydro-power can also be operated in a similar fashion. One problem is that it is always a possibility that this reserve fails to start up quickly enough.
- Spinning reserve: this reserve comes from the plant that is connected to the system, but which is operating below its peak load. The spinning reserve provides a quick response to demand changes and is the most reliable way to provide reserve. It does, however, involve a certain cost in maintaining it, not incurred with a quick start reserve.

2.5 System Balance

Because of the above conditions, decisions have to be made about what type of plant to be operated at any moment in time. These dispatch decisions need to take into account a variety of factors. The aim of dispatching is to decide how much each plant should generate electricity so that minimum operating costs are incurred for the system.

- A decision might be made to use all the plant to supply the load all of the time. This would mean that many plants would run at low loads and have poor thermal efficiency. Operating costs would also be high.
- Alternatively, the decision might be made to use those plant which meet the load and enable plant to operate at high levels of efficiency keeping operating costs low.

To achieve the latter, a merit order might be used where a priority list is created which ranks the generation units in some preference order. A common criterion used is the hourly fuel cost per megawatt (MW). A list would therefore be created ranking all generation units from the lowest to highest cost per MW hour, and units required to meet the demand would be selected on this basis. This approach is a simple one, and it is a static approach that assumes that operating costs do not change with plant output levels. It also does not take into account distribution and transmission constraints.

In making use of this ranking, it might be revised to take into account any factors such as start-up costs, minimum loading conditions, shut-down costs and minimum loading conditions. Normally, the dispatching decision will determine the short-run marginal cost of the system in each period. The short-run marginal cost is the operating cost of the most expensive plant that is being used to generate electricity in any given period. In a marginal cost-based pricing system, this is the cost that is used for setting prices.

In determining prices, system operators are generally mindful of any incremental costs. These costs are the additional fuel costs and other costs such as labour, materials and supplies incurred in increasing the supply of electricity. Often, it is more difficult to establish the level of the non-fuel incremental costs, but generally, fuel costs can be identified by relating them to the fuel input required for a plant at different output levels.

When using an incremental cost approach, the aim is to select units in the dispatching decision based on the operating costs alone. In doing so, a number of aspects related to plant operations are not considered, or given low priority in making the decision. The unit commitment is the decision-making about the economic scheduling of the generation units. In doing so, unit and system constraints need to be considered.

Unit constraints create restrictions on how a plant can be used. A plant, for instance, cannot be used beyond its maximum capacity. There is also a minimum load below which a plant will not operate. If a plant is suit down, it will take a minimum amount of time to return to production (the minimum downtime). The loading pattern of each plant also varies according to the ramp-up rate. Finally, there are costs associated with starting up a plant from scratch when it is cold and costs for shutting plant down as well. The impact of these constraints has to be costed and taken into account.

At the system level, the constraints need to be taken into account as well. These constraints will have an impact on more than a single generation unit. To maintain the system, there needs to be a significant reserve to face any emergencies. If regulation requires meeting certain environmental standards, then the system scheduling needs to take these into account as well. Transmission line capacity constraints can also be important.

As the industry is a capital-intensive one, there is a relatively long period for electricity industry projects to be constructed and brought into use. This means that decisions regarding additions to generation and transmission capacity are required well in advance of their being used. In addition, as there are a number of technologies to produce electricity, these long-term decisions need to choose which type of technology will be used, both now and into the future to meet demand. There is a large and well-developed body of literature that deals with the issues involved in long-term electricity resource planning that focus on the timing and optimal sizing of new capacity. For a review of this literature, see Hobbs (1995)and Foley et al. (2010). This resource planning can be considered by making use of an integrated approach where supply, reliability, demand, resource availability, demand-side options, financial issues and rates are considered together. The problem worth taking all of these into account is that they tend to be very demanding in terms of computational requirements and technical capabilities. An alternative to this is to use a modular approach where each module focuses on an individual issue and the results are then linked together to come to a final decision. As these decisions can have long-term impacts on the supply of electricity, and because demand is often changing quite rapidly over time, it is not to be unexpected that mistakes are made in terms of the types and size of capacity built. This has certainly occurred in the Chinese case, which will become clear from the work in this book.

2.6 Conclusion

In this chapter, an introduction to analyse of the decisions on creating the electricity supply has been provided. In doing so, the concept of load and energy demand and their influence on the supply technologies used have been provided. The chapter has then presented the challenges that face markets both in terms of operating decisions and the planning of investment. The purpose is to provide the reader with the knowledge of some of the basic problems that face any country in building up a suitable electricity supply industry. In order to appreciate these problems, it is necessary to be familiar with such terms as a merit order dispatch and priority list for unit commitments.

It is now possible to go and look at some of the more specific problems that face the Chinese electricity industry. As can be appreciated from the material presented in this Chapter, the decisions around the types of fuels used and its location can have a critical impact on other parts of the industry because of the interconnection between them. This has very special implications for the way in which the Chinese industry has developed in the past and how it will develop in the future.

References

BP. (2019). *BP statistical review of worked energy*, London: BP.

Foley, A. M., O'Gallachoir, B. P., Hur, J., Baldick, R., & McKeogh, E. J. (2010). A strategic review of electricity system models. *Energy, 35*(12), 4522–4530.

Hobbs, B. (1995). Optimization methods for electric utility resource planning. *European Journal of Operational Research, 83*(1), 1–20.

United States, Department of Energy, Energy Information Agency. (2020a). *Annual energy outlook 2019*. Washington: EIA.

United States, Department of Energy, Energy Information Agency. (2020b). *Country analysis briefs: China*. Washington: EIA, May.

Chapter 3
The Historical Construction of the Electricity Industry in China

3.1 Introduction

Because of the rapid development of China demand for electricity has grown at a very fast pace, which, in turn, has meant that the Chinese Government has faced constant pressure over a long period to maintain a consistent balance between electricity supply and demand. At times, the impetus for reform in China has been the existent of imbalances in the form of either shortages of supply or and overextension of the construction of new facilities. A growing shortage of electricity in China in 1985, for instance, forced authorities to first make changes in electricity utility investment and pricing policies. Since then, the electricity industry in China has gradually moved from being a strictly state-owned one to an industry characterised by investment from various sources, these including private enterprises, local governments and collective groups. Prices have also been modified from being strict state-controlled ones, to markets with competitive wholesale prices (with retail prices still largely controlled).

Although the literature on electricity markets suggests that these types of governance choices can have potentially large effects on the development and efficiency of this industry, in the Chinese case there have been few attempts to quantify the effects of these major changes in China's policy and governance practices on performance. In the future, it is expected that a growing body of work will be undertaken on these issues.

Investments in the electricity utility industry can be characterised by their asset specificity. They have large sunk costs, which can give rise of regulatory opportunism. That is, once a utility company has made a sunk investment in plant, it is vulnerable to being pressured by government regulators to keep prices low. In addition, regulated companies might try to opportunistically undertake excessive investments at inflated prices, as they often are in possession of information that is not known to regulators. How regulation is designed, therefore, can have an important impact on investment and on the levels of efficiency achieved. Heavy sunk costs also mean that past investment decisions can impact the industry over a very long time frame.

M. Xiaoying and M. Abbott, *China's Electricity Industry*,
SpringerBriefs in Energy, https://doi.org/10.1007/978-3-030-53959-7_3

In the Chinese case, much of the structure of the industry is influenced by these long past investments and responses to changes in regulation. Understanding the basic character of the industry over the longer term, therefore, it is important in understanding the present-day state of the industry. In this chapter, therefore, the focus is on the past performance of the industry in China and the manner in which it has evolved over the long term. In doing so, it looks at some of the past electricity market reforms and their effect on efficiency.

3.2 Background

To date, a number of studies have been undertaken that provide an examination of the Chinese electricity industry's long-term development. These studies can be divided up into three broad groups. The first group of studies tend to focus on the characteristics of past electricity market reforms and the levels of efficiency in the industry. Ma and He (2008), for instance, characterised the electricity market reforms as falling into five distinct phases. Andrews-Speed et al. (2000), on the other hand, divided development of the industry into four phases, concentrating on the Guangdong Province. In addition, Andrews-Speed and Dow (2000) emphasised that China's electricity industry is larger and more complex than that of other developing countries.[1]

The second group of studies focuses on the relationship between electricity industry development and economic growth. These studies have found a causality that runs from electricity consumption to growth in real GDP, but not vice versa, in the years 1971–2000. During the 1990s, rapid growth in electricity demand from the residential, commercial and industrial sectors also contributed substantially to shortages, but supply-side factors were more important, particularly, the construction of new plant in response to demand growth.

The third group of studies on China's electricity industry tends to focus more on environmental factors such as the emissions associated with the industry's operation. One of the key conclusions made in these studies is that high levels of greenhouse gas emissions in China from the electricity industry are caused by the high use of coal, as well as the relatively low efficiencies of production from resource extraction, generation, transmission and final energy consumption.[2]

Past studies of Chinese electricity and industry development highlight the importance of electricity generation capacity for the country's economic growth. This industry development, in turn, has gone through two major stages. In the first stage, between 1949 and 1984, the industry (like other energy industries in China) was treated by the Chinese Government as simply being a means to an end, that is, as a subordinate sector whose aim it was to support the development of other industrial sectors. The performance of the industry itself in terms of technical and managerial

---

[1] Andrews-Speed and Dow (2000).

[2] Ma and He (2008) and Andrews-Speed and Dow (2000).

efficiency received little attention. To achieve this goal of support to other industries, a highly centralised administrative approach was regarded as appropriate. This approach, however, was to have long-term implications.

China's first power generation with an installed capacity of 11.67 kW (kilowatts) was launched in 1882 in the city of Shanghai. China built a total generation capacity of merely 1.85 GW (gigawatts) nationwide during the first 70 years from the launching of the first commercial power generator to the founding of the People's Republic of China in 1949.[3] The Chinese electricity generation capacity has been expanding at an accelerated speed since 1949. It took almost 30 years, from 1949 to 1978, for China to increase its total capacity to 57 GW. In contrast, the annual addition of generation capacity increased from a few GW in the 1980s to 10–20 GW in the 1990s. Installed capacity roughly doubled between 2003 and 2010, from 380 GW to 960 GW.[4] In 2006 alone, 102 GW of new generation capacity was added, an increment substantially larger than the UK's entire electricity system.[5] By the end of 2010, installed generation capacity of China's electricity industry was 960 GW, second only to the USA' roughly 1075 GW. Within the installed capacity of 960 GW, coal-fired power plants accounted for 74%, hydro-electricity for about 21%, while nuclear power, wind power and other energy types for 5%.[6]

The other important aspect of this growth in capacity is that China's energy sources tend to be geographically mismatched with demand, with supplies of coal being concentrated in the north of the country, hydro-sources concentrated in the southwest and middle of the country, and only nuclear energy being concentrated in the high usage regions. Construction and development of electricity transmission grids in China have tended to be slow. The seven individual grid systems function reasonably well in themselves, but interconnections not so much, resulting in the full potential on inter-regional flows not being realised.[7] At present, China lacks a unified electricity grid network across the whole country, although there are plans to establish one in the future.[8]

3.3 Socialist Development, 1949–1984

Although the foundation of the Chinese electricity industry predates the Communist Revolution in 1949 (see Picture 3.1 of a power station in pre-war Shanghai) much of the industry structure of it was largely formed in the period leading up to the 1980s. During the socialist period of development from 1949 to the early 1980s, four main phases can be identified. The first phase took place in the 1950s, when

[3] Tian (2008).

[4] China Electricity Council (2011).

[5] Steinfeld et al. (2009).

[6] China Electricity Council (2011).

[7] Fenby and Qu (2008) and Huang (2008).

[8] Wang et al. (2009).

Picture 3.1 Shanghai Power Plant in Yangshupu, 1928. *Source* Published in Hawks Pott (1930)

the Soviet Union had a major influence on China's development. This was a period of highly centralised administrative mechanisms, however, large-scale investment in the industrial sector and electricity capacity meant that electricity generation levels rose steadily. The second phase was the disastrous Great Leap Forward period, from 1958 to 1966. At this time, the Chinese Government tried to create large increases in iron and steel production, which, in turn, required substantial increases in energy consumption. As it turned out the planned increases in steel production were never realised and so the demand for electricity anticipated was also not forthcoming. The third main phase was that of the Cultural Revolution from 1966 to 1976. During these two periods, China's economic development largely stagnated, and energy demand only increased slowly. The final period was from 1978 onwards when the country began to experience some market reform and opening up to international trade, which meant that the economy began to develop and electricity supply shortages became apparent.

During the first stage of the industry, during the 1950s, electricity prices were held below-average costs, which made it difficult for the industry to finance its own development. The Chinese socialist economy at this time was organised into work units, which paid for the electricity used not only at work, but also in workers' housing. Electricity was, therefore, considered an entitlement, and as a result of the very low electricity prices, electricity was often used in a wasteful manner. This situation continued into the post-1950s period and by the mid-1980s, the shortage of electricity became increasingly serious. In the early 1980s, electricity shortages were having a detrimental effect on industry development and inhibited the development of living standards. Restrictions on residential use of electricity were used to maintain factory production. These sorts of problems put pressure on the Chinese Government to undertaken reforms of the industry.

3.4 Reform from 1985

The shortages of the early 1980s meant that the Chinese Government undertook a series of important electricity regulatory policy changes. In 1985, the policy documents entitled: the 'Interim Provision on Promotion Fund-Raising for Electricity Investment and Implementing Multiple Electricity Prices' and 'The Measures of Implementing Multiple Electricity Prices' were released. These two policies when implemented were successful in overcoming many of the shortages of electricity in China. In fact, by the mid-1990s, the Chinese economy was to become one of surplus in the supply of electricity. This was due in part to the rapid growth in the construction of electricity generation capacity initiated by the new investment policies, as well as the levelling off in demand for electricity demand caused by the Asian Financial Crisis in 1997. From 1996 to 2001, the Chinese Government resorted to a more strict control of investment in the industry, which again resulted in renewed electricity shortages from 2002 onwards.

This period of supply shortage turned out to be a relatively short-lived one and the breakup of the vertical monopoly held by the State Power Corporation carried out in 2002 was to lead to another boom in investment in electricity generation capacity. As mentioned in the first chapter of the book, in December 2002, the State Power Corporation was dismantled and five big independent power generation corporations were created. The subsequent competition between the five corporations accelerated investment in generation in new capacity from 2003 onwards, however, coal prices rose sharply in 2003, which, in turn, made new generation less economical and tempered this growth in capacity from 2004 onwards.

Through the 2000s, electricity supply shortage was tempered to some degree, however, the shortage was not completely eliminated because of a related shortage in the supply of coal. The shortage of coal supplies had a number of causes including, a lack of railway capacity which made it difficult to increase supplies, and a lack of a market mechanism for coal which meant that coal supplies were unresponsive to shifts in demand. Once coal market price reform occurred, the results tended to be uneven in impact as prices were reformed to change with demand, but as electricity prices were still largely fixed, electricity generation firms were unable to pass on the increased costs of coal to final users. Many coal-fired plants, therefore, kept production at levels below full capacity, causing shortages. In addition, as coal could increasingly be sold to other buyers on the spot market, there were incentives to limit sales to electricity generators, but instead divert it to other industries.

Overall China's original centralised electricity management system led to severe problems in the allocation of energy resources. This meant that by 1985, shortages encouraged various reforms to occur that helped to bring forth new generation investment and capacity. Continued price controls of final electricity prices, and of coal supplies, however, led to continued distortions in markets, which, in turn, led to periodic shortages of generation capacity. Chinese authorities were also gradually to recognise that price controls in coal markets would need to be reflected in the electricity market. The resultant distortions eventually led to pressure to build up for additional reforms to be implemented.

3.5 Electricity Market Reform and Efficiency

Eventually, as reform has progressed the emphasis of the Chinese Government has changed from a desire to promote investment in generation capacity towards a need to introduce market-oriented mechanisms that encourage an improvement in the efficiency of the operation of the electricity industry. The emphasis in this process, besides structural change, has been to reform pricing.

As part of this process in 1993, the Chinese Government adjusted the catalogue price for electricity from six to eight groups. The purpose of this change was to help to balance supply and demand and also to enable commercial businesses to make better use of electricity. At the same time, residential electricity prices were much lower than industrial and commercial price (subsidies were given for residential electricity), and therefore often residential houses or apartments were used for commercial purposes to take advantage of this. The changes were designed in part to give less encouragement to this practice.

An additional reform was a change in pricing for high energy-intensive industries. One of the main aspects of economic development in China from the 1980s onwards was the development of energy-intensive industrial sectors. In order to lower the excessive increase in demand for electricity, energy-intensive sectors were increasingly charged higher prices. The major associated document for this measure was the: 'Notice on Further Implementation of Discriminative Power Price and Charges on Self-owned Power Plant', which aimed to reduce the growth rate of high energy-intensive sectors by increasing electricity prices.

These new prices tended to be a type of cost-plus regulatory prices (or rate of return prices), where profits were guaranteed and therefore the investment in generation plant was encouraged. The opportunity to overstate costs on the generation plant, however, did take place in order to get higher prices and revenue. It also led to the construction of electricity plant in this period that was high in cost and had low levels of energy efficiency. Investors at this time chose to construct small units, as these units required less capital and had simpler approval processes. They did, however, typically had lower economies of scale and therefore lower efficiency levels. At the end of 1996, the average thermal power unit capacity of the Chinese industry was only 46 MW (according to the China Electricity Council). Later as policies

Picture 3.2 Part of the Three Gorges Project. *Source* Authors' collection

were introduced to encourage the creation of larger scale units the average unit size increased and by the end of 2009, thermal power capacity units with 300 MW accounted for approximately 70% of total generation capacity.[9]

As well as investment in coal-fired plant large-scale hydro-schemes was invested in. The construction of the Ertan Hydro-Electricity Project was started in 1991 and was completed in 2000. Investors in this project were the State Development Investment Corporation, the Sichuan Investment Group Corporation, and the Sichuan Electric Power Corporation. The other most well-known Chinese hydro-electricity project was the Three Gorges Project (the largest in the world; see Picture 3.2). A substantial part of the investment in the Three Gorges Project came from the public through the issue of bonds and stock.

The new electricity generation units built after 2002, therefore, focused on larger scale generation units. This larger scale plant was a major contributor to China's electricity reform achieving improvements in cost efficiency, as well as contributing to the creation of new plant to meet increasing demand. Regarding the oversight of the industry and prices charged the Chinese Government began to carry out yardstick price regulation from 2004 onwards. This involved comparing the different electricity units in terms of costs and prices against one another. The major aim of this change

[9]China Electricity Council (2006, 2017).

was to identify the least cost-efficient plant and to put increased pressured on this pant to improve their performance of shut down.

Despite all these efforts to increase capacity and to do so with greater levels of efficiency, a number of problems still remained. The first was the nature of the high-voltage transmission grid in China. China's coal mines are mainly located in the western regions of the country, while the main electricity load centres are in the eastern, coastal regions. To reduce energy loss during the transmission of electricity between these regions, therefore, the State Grid Corporation started in 2004 to consider the construction of an extra-high-voltage grid network. It was later planned that more than 400 billion RMB Yuan investment would be spent to construct the super-grid network within 15 years in China.

To help enable this to occur structural reform of the national electricity grid saw it grouped into six regional grids: north-east, south, east, north, east, north-west and central. Among these regional grids, electricity shortages have persisted in the north, south, north China, and north-east grids; with a balance in the east grid; while electricity surpluses were created in the north-east and north-west grids. Delivery of electricity across these grids from one to another has been limited by the barriers of grid structures, lack of interconnection capacity and exchange mechanisms. The construction of extra-high-voltage grid interconnections linking the different regional could potentially improves energy efficiency.

The other important technological change that is starting to occur in China is that of the introduction of non-hydro-renewable energy. In many countries around the world policies have been put into place aimed at reducing carbon dioxide emissions by promoting the use of non-hydro-renewable energy. In China, the government has acknowledged the importance of renewable energy for its future energy security and has created targets as well as put into place some policies that will help to achieve these. Further detail on some of these measures is provided later in this book.

3.6 Conclusion

The Chinese electricity industry has a long history dating back to well before 1949 and the Communist Revolution. The development of the industry today is affected to a large degree, by its past investments, which makes it difficult to change it quickly in response to contemporary imperatives. Large sunk costs in the development of coal-fired generation plant mean that the dependence on this fuel will exist for years. Dominance of the industry by state-owned firms is also likely to be one that lasts for many years to come as well.

The dominance of traditional fuels and state companies is encouraged by the nature of the electricity industry itself. The industry has monopoly elements within it, as well as a dependence on network infrastructure for its development as well as instantaneous use, all of which create special conditions for the industry compared to others. In understanding the future development of the industry, it is important to understand the nature of these elements.

References

Andrews-Speed, P., & Dow, S. (2000). Reform of China's electric power industry. Challenges facing the government. *Energy Policy, 28,* 335–347.
Andrews-Speed, P., Dow, S., & Gao, Z. (2000). The ongoing reforms to China's government and state sector: the case of the energy industry. *Journal of Contemporary China,* 9(23), 5–20.
China Electricity Council. (2006). *Bulletin of the national power industry statistics in 2005.* http://suo.im/2xqST.
China Electricity Council. (2017). *Bulletin of the national power industry statistics in 2016.* Beijing: CEC. http://suo.im/2xqST.
China Electricity Council. (2011). *China electric power yearbook 2010.* Beijing: CEC.
Fenby, J., & Qu, D. (2008). *China's grid power up.* London: Trusted Sources UK Ltd.
Hawks Pott, F. L. (introd.) (1930). *Shanghai of to-day. A souvenir album of fifty vandyke gravure prints of the 'model settlement',* Shanghai: Kelly and Walsh.
Huang, H. (2008). *The new takeover regulation in China: Evolution and enhancement* (Working Paper 42). University of New South Wales Faculty of Law Research Series, Sydney, Australia.
Ma, C., & He, L. (2008). From state monopoly to renewable portfolio: Restructuring China's electric utility. *Energy Policy, 36,* 1697–711.
Steinfeld, E. S., Lester, R. K., & Cunningham, E. A. (2009). Greener plants, grayer skies? A report from the front lines of china's energy sector. *Energy Policy, 37,* 1809–1824.
Tian, J. (2008). *Implementing energy efficiency programs in china's power generation sector: Case study of a recent policy initiative.Regional and Sustainable Development Department,* Asian Development Bank, Manila.
Wang, Q., Qiu, H.-N., & Kuang, Y. (2009). Market-driven energy pricing necessary to ensure China's power supply. *Energy Policy, 37,* 2498–2504.

Chapter 4
Structure and Reform

4.1 Introduction

Market reform of the electricity industry has a long history before it was introduced into China and has been an international trend since the 1980s. Globally, the market-oriented reform can be traced back to the Chile electricity reform in 1982. The Chilean model was generally perceived as a good example of bringing rationality and transparency to electricity pricing. The Chilean practice has influenced subsequent reforms in other countries' electricity industry. A landmark electricity reform took place in UK from 1989 onwards. The UK Government restructured and privatised the stated-owned Central Electricity Generating Board, in doing so separating the ownership and operation of generation from distribution and transmission. The experiences of British were then used as a model or a catalyst for the deregulation of other countries' electricity industries.[1]

Meanwhile, it also should be pointed out that the market deregulation in many of other countries occurred without the widespread privatisation that characterised the UK model. For example, Norway's electricity reform has maintained a dominantly public ownership and decentralised production structure.[2] Although the institutions and market designs and deregulation procedures tended to be very different; the two principal underlying concepts are the same. These are:

- Separating the contestable functions of generation and retail from the natural monopoly functions of distribution and transmission; and
- Establishing a wholesale electricity market and a retail electricity market.

The role of the wholesale market is to allow trading between generators, retailers and other financial intermediaries both for short-term delivery of electricity and for future delivery periods.[3]

---

[1]Thomas (2006).

[2]Midttun and Thomas (1998).

[3]Sioshansi and Pfaffenberger (2006) and Woo et al. (2003).

M. Xiaoying and M. Abbott, *China's Electricity Industry*,
SpringerBriefs in Energy, https://doi.org/10.1007/978-3-030-53959-7_4

In this chapter, we focus on discussion of market-oriented reform in the Chinese electricity industry. The chapter is organised as follows. Section 4.2 investigates China's electricity reform from an absolute to relative monopoly. Section 4.3 discusses the relationship between the electricity and natural gas industries. Some conclusions are provided in the last section.

4.2 China's Electricity Reform

In terms of the change in the structure of China's electricity utilities, the market-oriented reform can be divided into three phases. Before the economic reform and the opening up of the country to the world economy in 1978, all electricity generation, distribution and transmission assets were state-owned. China has gradually started deregulating and de-centralising the economy from 1978; however, the electricity industry was one of the last sectors to be reformed and deregulated.[4]

The administration, resources allocation, investment decisions and pricing in the electricity industry were all fully controlled by the central government before 1985. Because all electricity assets were organised as a vertically integrated, state-owned utility, all prices were simply internal transfer prices. Electricity prices were, therefore, used virtually for accounting purposes rather than for the allocation of resources. On the retail side, prices were guided by so-called catalogue prices. These prices were deliberately kept low to support economic growth. In addition, only investment from the central government was able to be approved under the planned economy model of the time. Insufficient investment in the generation and transmission capacity was a chronic problem and the rapid growth in electricity demand driven by economic growth after 1978 helped to result in power shortages and black outs.

The shortage of electricity became a serious bottleneck holding up economic development between 1978 and 1985 and created pressure for reform to occur.[5] Eventually, the Chinese Government allowed for domestic private enterprises and foreign investors to invest in the generator sector from 1985 onwards. This change, however, did not apply to investments in the distribution and transmission sectors which were still forbidden. This policy strongly promoted the construction of additional electrical capacity, and thus increased the supply of electricity. By 1997, the nationwide chronic power shortage had been by and large eliminated.[6]

The reforms of 1985, however, created new challenge, especially in terms of electricity generation tariffs. The generation tariffs approved by the State Planning Commission and its successor, the National Development and Reform Commission, were designed to cover not only capital costs and loan interest, but also the cost of fuel, and transport. Such generation tariffs guaranteed the investors that their investment would be recovered over a repayment period determined by the Government,

[4]Wang and Chen (2012).

[5]Zhou (1999).

[6]Xu and Chen (2006) and Wang et al. (2009).

generally within ten years. This rule of the: 'repayment of principal and interest' resulted in a distorted electric pricing system that is greater revenues the greater the cost of generation was. Even worse, efficiency improvements and technology innovation were effectively discouraged by such energy pricing.[7] Finally, as local government obtained more jurisdictions over the development of the local electricity industry, local protectionism becomes common, which created institutional barriers for inter-regional trading of electricity and efficient resources allocation.[8]

Eventually, the Chinese Government decided to take up the task of reform again, and in March 1997, the State Power Corporation was established. In 1998, the Ministry of Electric Power was dissolved and its administrative functions were transferred to a new department under the State Economic and Trade Commission. The newly founded State Power Corporation, however, controlled around one half of the country's generation assets and almost all of the country's distribution and transmission assets. As a result, the State Power Corporation itself became a key obstacle to market-oriented reform of the Chinese electricity industry.[9]

In 2002, the Scheme for the Reform of Power Industry (the Scheme) was enacted following the examples mentioned earlier of international deregulation practices.[10] The overall objective of the scheme was to develop the electricity market system with the characteristics of the separation of government and business functions, and the promotion of fair competition. The main measures involved the breaking up of the state monopoly and the advancement of energy pricing mechanisms.[11]

The State Power Corporation was, therefore, taken apart as a vertically integrated utility. This occurred by most of its generation assets being reallocated to five large generation corporations, and its distribution and transmission assets being allocated to two transmission companies. At the same time, four consultant and construction companies were also split from State Power in order to enable them to compete with each other, and hopefully therefore to achieve higher levels of efficiency. Meanwhile, a new ministerial-level industry regulatory agency was created, the China Electricity Regulatory Commission, which was tasked with the role of advancing electricity price reform.[12] Taking up this role was not a smooth one, as the National Development and Reform Commission refused to yield key tools to State Electricity Regulatory Commission, including the authority to determine prices and approve new capacity installations. This refusal prevented State Electricity Regulatory Commission from fulfilling its mission.[13]

---

[7] Wang (2007). This is effectively what is known in economics as the Averch-Johnson affect.

[8] Ma and He (2008).

[9] Ma and He (2008).

[10] Ma and He (2008), Zhao (2004) and Wang and Chen (2012).

[11] State Council of China (2002).

[12] Xu and Chen (2006), 'The Reform of Electricity Power Sector in the PR of China'. Wang and Chen (2012), 'Regulatory failures for nuclear safety'.

[13] Wang et al. (2009), 'Market-driven energy pricing necessary to ensure China's power supply'. Wang and Chen (2012), 'Regulatory failures for nuclear safety'.

Serious power shortage during 2003 and 2004 helped to stifle the 2002 reform. Under the pressure of power shortages, expanding generation capacity and ensuring the supply of electricity took priority over the introduction of further market-oriented reforms.[14] In 2004, 24 provincial regions China experienced power shortages, with the deficit amounting to around 10% of installed capacity.[15] These shortages helped to stimulate expansion of the construction of generation capacity. The five generation companies took the lead in expanding the construction of generation capacity and from 2003 to 2010, installed capacity of China Huadian Corporation, China Guodian Corporation, China Datang Corporation, China Huaneng Group and China Power Investment Corporation have increased by over 300% in the case of each company. China's total installed capacity is increased by 250%. This meant that the share of capacity held by these companies rose over the period and was around 50% in 2010.[16] Other central state-owned generators controlled around 11% of capacity.[17] These companies were all owned by a single holding company, the state-owned Assets Supervision and Administration Commission. Before Scheme 2002, the central state-owned power generator, i.e. the State Power Corporation, controlled 46% of the total installed capacity. In one sense, therefore, the former monopoly power held by the state utility sector was enhanced by the reform of 2002.[18] At this stage, there was still little competition in the generator electricity market, with competition mechanisms only being simulated in pilot markets.[19]

The monopoly control of generation had a number of side effects, besides the encouragement of inefficient generation. One example was the case of wind power and solar energy. To stimulate the development of wind power, a Concession Project Bidding policy has been adopted in China since 2003. According to the rules of the Concession Project Bidding, the lowest bid usually wins. The big state-owned power generator, especially the big five were often prepared to foregone revenue, sometimes willing to shoulder losses in order to bid the lowest price. As a result, the winning bid price is so low that a reasonable profit would not be guaranteed. This, therefore, became an investment barrier for both private and foreign companies. In the Concession Project Bidding, some privately owned companies ventured into the price competition in the first few rounds.

In subsequent years, most private companies have declined to bid due to the high investment risks. Foreign companies have also decided to stay out of the bidding process. The state-owned companies can afford to lose money as they are often get subsidies for their losses. For private companies and foreign-owned companies, the risks are great enough to deter them.[20] A similar case applies to the solar photo-voltaic (PV) concession project. The bidders are almost all state-owned enterprises

[14]Wang et al. (2009).

[15]Xin (2004).

[16]State Electricity Regulatory Commission (2009, 2010, 2011).

[17]State Electricity Regulatory Commission (2011).

[18]Yeh and Lewis (2004).

[19]Ma and He (2008).

[20]Li et al. (2008) and Wang (2010).

Fig. 4.1 Natural gas proportion of total energy use in China, 1965–2018 (%). *Source* BP (2019)

as they are the bidders who can propose the lowest ranges of tariffs. Although such competition can drive an increase in wind power and PV electricity, it is unclear if it is beneficial for the future long-term development of the Chinese wind power and PV industry.[21]

The situation as it is in 2020 is as follows. China's electricity generation sector is still controlled by five state-owned generation companies—China Huaneng Group, China Datang Corporation, China Huadian Corporation, China Guodian Corporation and China Power Investment Corporation. These five companies generate nearly half of China's electricity (see Picture 4.1). Much of the remainder is generated by local-owned enterprises or by independent power producers (IPPs), often in partnership with privately listed arms of the state-owned companies. Deregulation and other reforms have opened the electricity industry to foreign investment, although investments have been limited so far.[22] The electricity distribution and transmission assets are still divided into two companies, the China Southern Power Grid Company and the State Grid Corporation of China, which operate the nation's seven power grids. The State Grid Corporation operates power transmission grids in the north and central regions, while China Southern Power Grid Company handles those in the south.[23]

[21] Wang and Chen (2010).

[22] IHS Energy (2014) and Bloomberg (2015).

[23] IHS Energy (2014).

The regulation enforcement of the electricity industry and facilitation of investment and competition to alleviate power shortages and today lodged in the NEA. On-grid (electricity sold by generators to the grid) and retail electricity prices are determined and capped by the NDRC. The NDRC also determines the price that coal companies should receive from power producers for a certain level of electricity. Policy today is mainly seeking to improve system efficiency and the interconnections between the grids through ultra-high-voltage lines, as well as to implement a smart grid plan. The first phase was completed in 2012, and subsequent phases are slated for completion in 2020.[24]

Before 1985, the Chinese electricity industry was an absolute monopoly. This absolute monopoly was broken in 1985, when the Chinese Government allowed the local government, domestic enterprises, foreign and private investors to invest in the generation sector in order to overcome electricity shortages. In 2002, the Scheme for the Reform of the Power Industry was enacted, which was aimed at developing a more market-oriented electricity system. Electricity shortages in 2003 and 2004, however, made expansion of generation capacity more important than market-oriented reform. Since then the state-owned enterprises have retained, and in some ways, enhanced their positions. The big five still control half of the national generation capacity.

4.3 Natural Gas and Electricity

Despite the continued control of much of the country's generation capacity there has been a gradual change in the composition of fuel sources. Since 2000, for instance, there has been in China a number of major new investments in natural gas infrastructure, which have gone a long way to developing natural gas as major energy source in China. In particular, work was completed on the construction of a 4250 km long west-east pipeline to supply natural gas to northern cities in China from the western region of Xinjiang, pipelines connecting China with Central Asian countries and Myanmar and the construction of major import terminals to supply southern China with liquefied natural gas (LNG) (Tables 4.1, 4.2, 4.3).

Historically, natural gas has made only a tiny contribution to the satisfaction of China's total energy demand. In recent years, it has become the national policy of the Chinese Government to promote the use of natural gas from sources both internal and external to the country. This is being done for a range of reasons, including the need to develop new sources of energy to meet expected future demand growth, to diversify sources of energy away from an over reliance on Middle Eastern oil and Chinese coal and finally to promote the use of fuels that have cleaner emissions. Emission levels of CO_2 are less than one half of most types of coal and considerably lower than that of petrol or diesel (Table 4.4).

In order to enable growth of natural gas use in China, a considerable amount of investment has needed to be undertaken in the development of natural gas resources,

[24]IHS Energy (2014).

Table 4.1 Population, energy consumption and average energy consumption, various countries, 2018, 2020 (number, millions tonnes oil equivalent, tonnes per person)

	Population (millions)	Energy consumption (million tonnes oil equivalent)	Average (tonnes per person)
China	1394	3274	2.3
USA	333	2301	6.9
World	7684	13,865	1.8
Japan	125	454	3.6
Taiwan	27	118	4.4
South Korea	52	301	5.8
China HK	7	31	4.4
Philippines	109	47	0.4
Malaysia	33	99	3.0

Source Central Intelligence Agency (2020), BP (2019)

Table 4.2 Imports of natural gas into China, 2018 (billion cubic metres)

Pipeline (*Total* 47.9)	
Kazakhstan	5.4
Turkmenistan	33.3
Uzbekistan	6.3
Myanmar	2.9
LNG (*Total* 73.5)	
Qatar	12.7
Australia	32.1
Malaysia	7.9
Indonesia	6.7
Nigeria	1.5
Europe	1.2
New Guinea	3.3
Angola	0.7
Trinidad and Tobago	0.5
Russia	1.3
Total imports	120.4
Total Chinese consumption	283.0

Source Central Intelligence Agency (2020) , BP (2019)

Table 4.3 Composition of electricity generation, China and various countries, 2018 (%)

	China	Japan	Korea	Taiwan	USA
Coal	67	33	44	46	28
Hydro	8	8	–	2	6
Natural gas	3	37	27	35	35
Oil	–	16	2	3	1
Nuclear	4	5	22	10	19
Renewables	9	11	4	2	10

Source Federation of Electric Power Companies of Japan (2014), Central Intelligence Agency (2020) and BP (2019)

Table 4.4 lb of CO_2 emitted per million BTU of energy for various fuels, USA (pounds/million BTU)

Coal (anthracite)	228.6
Coal (bituminous)	205.7
Coal (lignite)	215.4
Coal (sub-bituminous)	214.3
Diesel fuel and heating oil	161.3
Petrol	157.2
Propane	139.0
Natural gas	117.0

Source United States, Department of Energy, Energy Information Agency (2020)

along with associated infrastructure such as treatment plant, storage facilities, import terminals and distribution and transmission pipelines. The sheer scale of the necessary investment needed in infrastructure has meant that the Chinese Government has begun actively to encourage foreign investment in an industry that until recently was dominated exclusively by Chinese Government companies. This desire on the part of the Chinese Government to attract foreign investment into natural gas infrastructure has meant that the Government has had to establish a legal and regulatory regime that gives foreign investors a degree of security over their investments.

The natural gas industry, similar to the case of electricity, is typically divided into broad sub-sectors corresponding to the three distinct sets of physical operations. These three sectors are natural gas production, distribution and transmission.[25] Production encompasses the exploration and development of the natural gas reserves. Traditionally, in most countries, this aspect of the industry grew out of the activities of the oil companies (whether privately or government-owned), which in the search for crude oil discovered natural gas together in the same geographical structures. The distributors of gas, on the other hand, are generally the direct successors of companies, which first illuminated streets with gas manufactured from coal or

[25]Gianna (2011) and Tussing et al. (1995).

oil and subsequently supplied homes and businesses with this type of gas. These companies were often owned by either national or local governments. The natural gas transmission sector in most countries evolved from the efforts of oil companies, or combinations of distributors and oil companies, which sought to link local gas utility grids to the natural gas reserves.

The structure of the natural gas industry in most countries has been determined by the separate origins of the three fundamental segments of the natural gas industry. As well it has been influenced to a considerable degree by the manner in which respective governments regulate the industry. The structure of the natural gas industry in the various countries generally follows either of three main models (vertically integrated; separated; market-based).

(1) Vertically integrated: in a number of countries, the distribution and transmission segments of the industry are bundled together into a vertically integrated monopoly provider of natural gas. Examples of this structure are the state-owned companies in France (GDF), Indonesia (Pertamina), Korea (Kogas) and Malaysia (Petronas).[26] Often these companies are also involved in the upstream production and exploration segments of the industry. This structure is a common one traditionally where the national government of a country has taken a conscious decision to take over the industry and operate it as a vertically integrated entity. As well as the monopoly provision of gas vertically integrated companies often have responsibilities over the regulatory aspects of the industry (such as planning approval, environmental, safety, quality standards and pricing). In more recent years, there has been a tendency to concentrate the activities of these state-owned companies on commercial activities and to establish more formal regulatory functions attached to separate government authorities.
(2) Separated: a common structure where the final sale of gas to consumers is dominated by privately owned utilities or local government-owned distributors. With this model gas producers sell gas under long-term contracts to transmission companies who then on sell it to distributors, who in turn retail the gas to final consumers. Traditionally, consumers purchase gas delivered by a single monopoly distribution network owner at a bundled price incorporating the cost of the gas plus the cost of transportation. The distributors generally have a legal or de facto exclusive monopoly right to sell gas in their area to consumers in return for an obligation to supply in those areas.[27] Transmission operators might sell directly to large industrial consumers of power generators. This model exists in countries like Germany, Italy, the Netherlands, Belgium and Japan. Given the monopoly nature of the distribution and transmission contract prices under this model are generally regulated by the government either by a formal regulatory agency or under informal arrangements. Other regulatory arrangements

[26] ABARE/Asia-Pacific Economic Cooperation Energy Working Group (2002).

[27] Horizontal integration of gas distribution has occurred in number of countries where by gas is distributed along with other services such as electricity, telephone and cable television.

might be divided between different levels of government; for instance, distribution pipeline planning approvals and retail pricing might be a local or regional government responsibility and transmission planning and contracts a national government responsibility. Given the lack of competition in this model, which can lead to the achievement of a less than optimal level of economic efficiency, a more market-based approach has been developed in a number of countries.[28]

(3) Market-based: in the recent reforms of the natural gas industry in a number of countries, a central issue has been the creation of more competitive and flexible markets. Gas reform, therefore, has generally involved the unbundling of the market for the gas commodity from the market for gas transport services (i.e. access to distribution and transmission transport services). The purpose is to separate and regulate the natural monopoly part of the supply chain (the transport services) and allow competition to develop in the potentially competitive sections of the industry; gas production and supply. Unbundling and open access to pipelines under this model create the opportunity for gas consumers and producers to negotiate directly for the sale of gas and then transport it separately. Vertical separation between transmission, distribution and supply is generally carried out in this process. Price regulation is generally concentrated on the natural monopoly components of distribution and transmission pipeline access. The regulatory structure under this model generally requires the establishment of an access code (either voluntarily or through government mandate) as well as protocols for the switching of customers from one supplier to another. This form of separation and open access has occurred in Australia, Canada, Chile, Mexico, New Zealand and the USA and is being considered or proposed in Indonesia, Thailand, Japan, South Korea, a number of Western European countries and Singapore.[29]

Government intervention exists under each model although the nature, scope and intensity vary depending on the model used. The form of government intervention might involve government legislation, state ownership, rules applied by regulatory bodies and taxation exemptions/royalties. The reasons for this intervention have been numerous and include the desire on the part of governments to ensure reliable and secure energy supplies. The technical and economic characteristics of some parts of the energy supply chain that make it efficient to have a single supplier have also encouraged government intervention in energy markets either in the form of economic regulation of private operators or alternatively outright government ownership. Social and other policy objectives, such as ensuring the availability of energy at affordable prices for some consumers have also been important in some economies as have concerns about the environmental impacts on supply and use and safety standards. Often this intervention has evolved over time and in many countries lacks clarity and coherence.

[28] International Energy Agency (1998).

[29] ABARE/Asia-Pacific Economic Cooperation Energy Working Group (2002)and International Energy Agency (1998).

The form of regulation that applies to the industry is to some degree determined by the model of industry structure that exists. The first model, for instance, often has no formal price regulation, but instead some prices are determined by the state-owned company with possibly only some sort of general ministerial approval. In the second model, price regulation might be undertaken at a national level or alternatively at a joint national/local level. For final consumers, natural gas prices might be regulated by local or regional government authorities. National regulators might regulate the price of well-head gas or transmission prices. In the final model, typically, it is only access to the pipelines that are regulated. As the production and retails, sectors are the subject of competition regulation of prices at this level are generally considered unnecessary. Instead price regulation occurs only at the level of access to distribution and transmission network. Other aspects of regulation such as environmental, safety and planning approval processes are similarly affected by the structure of the industry.

In those countries, where the determination of pricing is formally regulated by a government agency either one of the three general approaches are used. The first is the cost-plus methodology where rates reflect costs in their pure form including the acquisition cost of the gas, labour, operation and maintenance plus a rate of return. This approach was used formerly in the USA and Canada and is still used in Japan and France for final consumers. Today, in the USA and Canada, it is the approach used to price distribution and transmission transport services. The second approach is the market value principle approach. Under this approach, the price of gas is determined as a reflection of the market value of alternative fuels (such as oil). The price paid by distributors is a commodity price including a calculated share of transportation costs. Final customers would pay a price based on the "market value" of the gas along with additional costs of transportation. This approach is used in a number of European countries. The final approach is the use of price caps, which typically place a cap on maximum prices. The formulae for defining price caps can vary. They are usually set as a function of the evolution of inflation and productivity thus forcing the company to reduce costs and increase its productivity. The Retail Price Index (RPI)-X (or Consumer Price Index (CPI)-X as it is known in some countries) formula as it is applied in the UK is an example of this approach. In the British case, the RPI-X price cap approach was used before 1997 to determine retail prices (with a provision for a pass through of fluctuating gas supply prices). In countries such as Australia and Britain, today, this approach is used to determine the price of access to gas pipelines. With this approach, prices are allowed to rise each year in line with the RPI or CPI with an adjustment downwards to represent expected productivity improvements (the X value).

The rapid growth of the Chinese economy over the past thirty years has meant that growth in energy demand has been strong. China is the country with the world's largest population (1.394 billion in 2020) and has a rapidly growing economy, which has driven the country's high overall energy demand.[30] Growth in energy use in China has averaged about 4% a year since 1980 and it is expected that future growth over the next twenty years will achieve similar levels.

[30]Central Intelligence Agency (2020).

This makes China the largest global energy consumer, a position it has possessed since 2011. Despite constituting such a large share of the world's energy use per capita consumption is relatively low, compared to most developed countries. In 2019–20, China's per capita energy use stood at 2.3 tonnes of oil equivalent per capita compared to an average for the USA of 6.9. This figure is, however, above the world's average (at 1.8 tonnes), which indicates that China's energy consumption has greatly narrowed the gap in recent years, and will in likelihood be ranked at a similar level to that of Japan (at 3.6 tonnes), and South Korea within a few years (5.8 million) (Tables 4.1 and 4.2).

Coal is by far the most dominant fuel used in China because of the availability of low cost domestic reserves. In 2019, coal accounted for 54% of the total commercially traded energy production followed by oil at 20%. Coal is also the major source of fuel for electricity generation in China where it accounts for just 63% of the total (Table 4.3). Historically, natural gas has not been a major fuel in China constituting only 5.6% of commercially traded fuel in 2013 and contributing 4% of electricity generation (Table 4.3). Internationally this is a very low figure, as gas constitutes around one-quarter of the world's fuel share.

Given, however, the increasing need to import oil, the availability of domestic reserves of natural gas (proven reserves have been estimated to be 3.5 trillion cubic metres; and the environmental benefits of using natural gas the Chinese Government has decided to make a major effort to expand the use of natural gas and plans that it should reach 10% of its energy consumption by 2025.[31] Given the very small contribution that natural gas makes to total fuel use in China there would appear to be significant potential of its expanded use.

In the past, natural gas was used in China largely as a feedstock for fertiliser and chemical plant, with little use for electricity generation or residential use. In the future, natural gas is expected to be used extensively in power generation, industrial, commercial and residential use. Consumption of natural gas increased over fourfold between 2000 and the year 2010 and has increased even more over the past ten years. The largest share of this has come from industry, but the shares taken from power generation and transportation are rising.[32]

Even with this substantial increase in the natural gas industry in China, it is expected that gas will remain a relatively minor component of China's energy industry. Projections predict that natural gas's share of total energy demand will have risen from 2% in the 1990s to around 8% by the 2020s (see Fig. 4.1).[33]

Natural gas production in China grew strongly in the 1960s and the first half of the 1970s after the discovery of large fields in Sichuan Province. This growth tended to peter out in the 1980s. Despite the extensive size of the Sichuan reserves, further gas reserves needed to be developed and linked to China's urban centres to enable further expansion of natural gas consumption. The increase in the use of natural gas came from a number of major sources. The first was from domestic sources, primarily

[31] Facts Global Energy (2015).

[32] Facts Global Energy, East of Suez (2015).

[33] International Energy Agency (2000) and Lan Quan and Keun-Wook Paik (1998).

Picture 4.1 Cooling towers from nuclear power stations in China

from the reserves in the Xinjiang Province in the west of the country, but also from other reserves including some that are offshore. The second source is from imports of both pipeline and liquefied natural gas, the latter being used to supply consumers in the southern regions of the country—particularly power generators—who are more isolated from domestic (and piped imports) sources.

The country's largest reserves of natural gas are located in western and northern China, necessitating a substantial investment in pipeline infrastructure to carry it to eastern cities. The reserves in the Xinjiang Province in the northwest of China are made up of three separate basins (the Tarim, Junggar and Turpan-Hami). A recent survey by China National Petroleum Corporation (CNPR) reports that the Xinjiang region holds 34% of China's gas reserves. China built a pipeline from gas deposits in the Western Xinjiang Province to Shanghai picking up gas in the Ordos Basin along the way. Shell was chosen in February 2002 as the lead firm for the project and Gazprom and ExxonMobil held significant stakes.

The Chinese Government announced a discovery of a major gas field at Sulige in the Ordos Basin in the Inner Mongolia Autonomous Region adjacent to the Changqing oilfield in 2001. Estimates cited in the trade press put the reserves in the range of 16–21 trillion cubic feet. A pipeline was completed in 1997 between the Ordos Basin and Beijing and a second pipeline was built for natural gas in Beijing, Tianjin and Hebei Province. A pipeline was completed in 2002 linking the Sebei natural gas field in the Qaidam Basin with consumers in the city of Lanzhou. Another linked gas deposits in Sichuan Province in the south-west to consumers in Hubei and Hunan Provinces of central China at an estimated cost of $600 million.

Offshore gas projects are also important. The Yacheng 13-1 field was developed in the mid-1990s and has been producing gas for Hong Kong and Hainan Island since 1996. The Chunxiao gas field in the East China Sea is being developed by China Star Petroleum and is becoming a significant producer.

In 1998, the Chinese Government reorganised the state-owned oil and gas assets into two vertically integrated firms—the China National Petroleum Corporation

(CNPC) and the China Petrochemical Corporation (Sinopec). Before the restructuring, CNPC had been engaged mainly in oil and gas exploration and production. SINOPEC, on the other hand, had been engaged in refining and distribution. In 1998, the Chinese Government ordered a swap of assets, which transferred some exploration and production assets to Sinopec and some refining and distribution assets to CNPC. This created two regionally focused firms CNPC in the north and west and Sinopec in the south.

CNPC is by far the most important company in China in terms of the development of the natural gas industry as it controls the oil and gas reserves of the Xinjiang Province. CNPC carried out initial public offerings by the sale of a minority interest in its subsidiary called PetroChina in Aril 2000 on the Hong Kong and New York stock exchanges. The public offering raised over $US 3 billion with BP the largest purchaser at 20% of the shares offered. China National Offshore Oil Corporation (CNOOC) held its initial public offering of a 27.5% stake in February 2001. Shell bought a large block of shares valued at around $US 200 million. PetroChina put up 50% of the capital for the west-east pipeline along with the previously mentioned foreign companies and Sinopec. Sinopec operates the Puguang natural gas field in Sichuan Province.

The other main firms include the China National Offshore Oil Corporation, which handles offshore exploration and production and the China National Star Petroleum Company, which was created in 1997. The former developed the Yacheng 13-1 field in the 1990s, which produces gas for Hong Kong and Hainan Island. The latter company is developing the Chunxiao gas field in the East China Sea.

China National Offshore Oil Corporation is also responsible for the development of the import of LNG into the Guangdong Province. In this province, it originally constructed six, 320 MW gas-fired power plants and converted existing oil-fired plant with a capacity of 1.8 GWs to LNG. In March 2001, it was announced that BP Amoco had been selected to build China's first LNG import terminal valued at $US 600 million located near the city of Guangdong. BP took 30% equity stake in the project with China National Offshore Oil Corporation holding 31% and the rest being held by local and Hong Kong firms. Construction of the terminal began in 2002 and operations began 2005. Additional terminals were built by the company in Fujian and Shanghai.

On the 8 August 2002, CNOOC announced that Australia LNG won the contract to supply Guangdong LNG for the next 25 years and CNOOC also engaged in discussions with BP to supply further LNG terminals from the Indonesian Tangguh gas field and from Qatar. By 2014, China had become the third-largest importer of LNG behind Japan and South Korea. China imported 27.5 billion cubic metres of LNG in 2014, around 10% of international trade in LNG (Table 4.2).[34]

China's first international natural pipeline connection, the Central Asian Gas Pipeline, transports natural gas through three parallel pipelines from Turkmenistan, Uzbekistan and Kazakhstan to the border in Western China. Its first and second phases began operating in 2010. Investment in additional capacity to transmit natural gas

[34]ABARECONOMICS/Energy Research Institute (2003).

from Central Asia is being built. In addition, the China–Myanmar pipeline began operations in 2013.

Despite the opening of international pipelines and import terminals the bulk of natural gas consumed in China comes from domestic sources. In 2014, for instance, 58.8 billion cubic metres of gas was imported (31.3 by pile and 27.5 bcm at import terminals) compared to a total consumed in China of 185.5 bcm.

At the distribution level, natural gas is delivered to final consumers by local government-owned gas network companies. Traditionally gas to residential customers in China was and is provided by local government-owned network companies that deliver gas manufactured from coal or oil. Those cities that are now connected to natural gas have retained local government ownership of the distribution networks and deliver natural gas purchased under contract from the pipeline companies.

In order to achieve the exploitation of Chinese natural gas reserves, a considerable investment needs to be made into China's natural gas infrastructure as well as the development of China's natural gas reserves themselves. Poor natural gas infrastructure remains a critical issue in China. Development of a comprehensive transportation and distribution network is a necessity in view of China's projected increased natural gas use. Pipeline construction and downstream projects are still lagging behind upstream developments.

One major hurdle for natural gas projects in China is the lack of a unified regulatory system, which deters foreign investment. Currently, natural gas prices are governed by a patch-work of local, regional and national regulations. The Chinese Government is in the process of drafting a new legal framework for the natural gas sector, which is an urgent priority to reassure foreign companies that there will be a stable regulatory framework. Gas pipelines are capital intensive and the investment in them is very long term with much of it sunk (i.e. it cannot be retrieved if the company leaves the market). Countries, therefore, have to compete in international markets to attract private investment in these long-term infrastructure projects. International investors look for commercially viable projects that match returns on investments with the levels of risk they perceive. The less transparent and predictable the regulatory regime, the higher the degree of risk and the higher the returns that will be demanded, which can be affected by the laws regulations and institutional structure of a country.[35]

Before 1990 China's public utilities and energy sector made no effort to attract foreign investment. This is in contrast to the rest of the economy, which had become the world's second-largest destination for direct investment. Official hesitation to open the sector manifested itself in heavy intervention, long complicated approval processes and the lack of institutional and legal framework comfortable for investors. Since 1990, there has been some loosening up. Recent energy market reforms, expanded capital markets have spurred both foreign direct investment and foreign portfolio financing in Chinese energy. The state-owned enterprises have begun to

[35] Blackman and Wu (1999) and Perugini (1999).

participate in and seek ventures with foreign investors much more frequently than before.

Energy consumption in China is still below that on an average per capita basis compared to overseas and so it is expected that energy demand in China will continue at a brisk pace. In addition, gas consumption in China is on average well below that of most countries overseas. In China, in 2019, natural gas contributed only 7% of total energy use compared to 30% in the USA, 22% in Japan and 16% in South Korea. At the very least, it would be expected that natural gas use will increase at least to levels comparable to countries and regions such as Japan and South Korea (countries that depend on LNG).

The structure of the Chinese gas industry will depend upon the degree to which the major Chinese oil and gas companies and foreign investors are allowed to expand their activities downstream from the development of the natural gas reserves. At the very least, it would appear that there will be a degree of vertical separation in that local distribution and retail networks will be operated by local government bodies and price regulated by local and regional bodies.

At the transmission level, it would appear that major developments will be undertaken by joint Chinese/foreign investor's ventures with the former being given a majority interest. This will mean that gas production companies will be involved in the transmission of natural gas and will then contract with major gas users (i.e. power companies).

In some circumstances, there is a strong separation between the supply and transmission of natural gas. For instance, the LNG contracts to supply southern China are being undertaken by foreign companies who supply LNG for a Chinese dominated consortium which operate the LNG terminals and then transmission. This gas, in turn, will be provided for electricity companies and local gas network companies. There is therefore a fair likelihood that the second model of industry structure will evolve in China with separated gas development, distribution and transmission/retail companies emerging and contracting with each other with monopoly franchise areas for the distribution networks.

In terms of the determination of pricing, the government will still play a particularly important role in the determination of prices along the transportation chain. This is not unusual under the separated model with it being common under this approach that governments are heavily involved in the determination of prices at all contract levels including well-head, transmission–distribution and retail levels. In a number of countries responsibilities for this price regulation are separated between different levels of government: i.e. well-head at the national level and retail at the local or regional government level.

Natural gas currently plays a minor role in overall power generation and accounted for only 43 gigawatts of installed capacity at the beginning of 2014. The government, however, plans to invest heavily in more power plants fuelled by natural gas. Despite the traditionally cheaper use of coal as a source of fuel, improved gas generation techniques have made it cheaper to produce electricity through the use of natural gas. Table 2.1 in Chap. 2 provides data on the comparative cost of generating electricity using various sources of energy. Using combined cycle, natural gas plant is competitive compared to the use of coal, as long as reserves of natural gas are readily available and if the necessary infrastructure to transmit natural gas over long distances is available. With the construction of China's natural gas, transmission network being developed this should increasingly become the case.

Compared to other countries such as the USA, Japan and South Korea, the use of natural gas to generate electricity is relatively low implying that there is considerable scope for expanded use once reserves are developed and infrastructure built.

4.4 Conclusion

Reform of China's electricity industry has been a long and drawn out one and much needs to be done to complete it. There is still in China a dominance of the industry by a small number of generation companies. The development of new models is important if higher levels of efficiency are to be achieved. In some ways, expanding capacity and the opening up of new sources of energy has been the greatest achievements of the industry. Continuing this development will require new sources of fuel.

In terms of the natural gas industry, the opening up of the industry in China to foreign investment and competition has come much later than in other areas of the Chinese economy, even behind that of electricity generation. Even today, the gas industry, like electricity generation and transmission, is dominated by few state-owned companies that incorporate a high degree of vertical integration and possess substantial regularly functions. This type of structure is not that dissimilar to those that existed in many other countries prior to the 1990s and the reform of the Chinese gas industry is taken place in line with that of a great number of other countries. The structure that appears to be emerging is one where there is some separation vertically between producers, transmitters and distributors but the open-access model that has been created in North America, the UK and Australia/New Zealand appears to be a long way off.

The separated models of the electricity and gas industries' structure are ones that are particularly government regulation intensive with transmitters and distributors/retailers possessing a considerable amount of market power. The further development, therefore, of regulatory structures and pricing rules would appear to be an area of considerable activity in China over the next few years. The nature and stability of such rules are increasingly important given the likelihood of considerable privately sourced infrastructure investment in both industries. A lack of clarity on regulatory rules can quite easily help to deter private investment; therefore, the need to develop these rules would appear to have a particularly high importance.

Despite these impediments to foreign and private investment, there has been considerable capital expenditure in the natural gas industry over the past twenty years. This has tended to raise the share of natural gas as a proportion of China's total energy use, and of the country's electricity generation. There does, however, appear to be a long way to go until comparable proportions to countries such as the USA, South Korea or Japan are reached.

References

ABARE/Asia-Pacific Economic Cooperation Energy Working Group. (2002). *Deregulating energy markets in APEC: Economic and sectoral impacts.* Canberra ACT: APEC Secretariat.

ABARECONOMICS/Energy Research Institute. (2003). *Natural gas in eastern China: The role of LNG.* ABARE Research Report 03.01, Canberra, ACT: Abareeconomics.

Blackman, A., & Wu, X. (1999). Foreign direct investment in China's power sector: Trends, benefits and barriers. *Energy Policy, 27,* 695–711.

Bloomberg. (2015). *China issues rules to reform electricity system.* NE21s.

BP. (2019). *BP statistical review of worked energy.* London: BP.

Central Intelligence Agency. (2020). *World factbook.* Washington, DC: CIA.

Facts Global Energy. (2015). *China oil and gas monthly.* London: FGE. Dec, 2014.

Facts Global Energy, East of Suez. (2015, June 26). *Gas databook: Asia Pacific natural gas and LNG,* London: FGE.

Federation of Electric Power Companies of Japan. (2014). *Electricity review Japan.* Tokyo: FEPC.

Gianna, B. (2011). *Investing in energy: A primer on the economics of the energy industry.* Princeton: Wiley.

IHS Energy. (2014, May). *Energy country profiles—China, 29*:37–38.

International Energy Agency.(1998, 7 December). *Natural gas distribution: Focus on western Europe,* Paris: IEA.

International Energy Agency. (2000). *China's worldwide quest for energy security.* Paris: IEA.

Li, J., Gao, H., Wang, Z., Ma, L., & Dong, L. (2008). *China wind power report-2008.* Beijing: China Environmental Science Press.

Ma, C., & He, L. (2008). From state monopoly to renewable portfolio: Restructuring China's electric utility. *Energy Policy, 36,* 1697–1711.

Midttun, A., & Thomas, S. (1998). Theoretical ambiguity and the weight of historical heritage: A comparative study of the British and Norwegian electricity liberalisation. *Energy Policy, 26,* 179–97.

Perugini, A. (1999, 9–10 November), *Market development, regulatory framework and financial needs; Financing means for natural gas-fired power generation and city gas infrastructure in China,* IEA Conference on Natural Gas: Beijing.

Quan, L., & Paik, K.-W. (1998). China natural gas reports. Xinhua Agency Beijing and Royal Institute of International Affairs, London: RIIA.

Sioshansi, F. P., & Pfaffenberger, W. (2006). *Electricity market reform: An international perspective.* Oxford: Elsevier Science.

State Council of China. (2002). *The scheme of the reform for power industry in: China,* Beijing: SCC (in Chinese).

State Electricity Regulatory Commission. (2009). *Electricity regulatory annual report 2008.* Beijing: SERC.

State Electricity Regulatory Commission. (2010). *Electricity regulatory annual report 2009.* Beijing: SERC.

State Electricity Regulatory Commission. (2011). *Electricity regulatory annual report 2010.* Beijing: SERC.

Thomas, S. (2006). The British model in Britain: Failing slowly. *Energy Policy, 34,* 583–600.

Tussing, Arlon, & Tippee, Bob (Eds.). (1995). *The natural gas industry: Evolution, structure and economics* (2nd ed.). Tulsa, Okla.: PennWell Books.

United States, Department of Energy, Energy Information Agency. (2020). *Annual energy outlook 2019.* Washington: EIA.

Wang, B. (2007). An imbalanced development of coal and electricity industries in China. *Energy Policy, 35,* 4959–4968.

Wang, Q. (2010). Effective policies for renewable energy—The example of China's wind power—Lessons for China's photovoltaic power. *Renewable and Sustainable Energy Reviews, 14,* 702–712.

Wang, Q., Qiu, H.-N., & Kuang, Y. (2009). Market-driven energy pricing necessary to ensure China's power supply. *Energy Policy, 37,* 2498–2504.
Wang, Q., & Chen, X. (2012). Regulatory failures for nuclear safety—The bad example of Japan—Implication for the rest of world. *Renewable and Sustainable Energy Reviews, 16,* 2610–2617.
Woo, C. K., Lloyd, D., & Tishler, A. (2003). Electricity market reform failures: UK, Norway, Alberta and California. *Energy Policy, 31,* 1103–1115.
Xin, H. (2004). *Severe energy shortage warned.* Beijing: Xinhua News Agency.
Xu, S., & Chen, W. (2006). The reform of electricity power sector in the PR of China. *Energy Policy, 34,* 2455–2465.
Yeh, E. T., & Lewis, J. I. (2004). State power and the logic of reform in China's electricity sector. *Pacific Affairs, 77,* 437–465.
Zhao, J. (2004). Reflection on the reform of China's electricity industry. *Energy of China, 26,* 8–12.
Zhou, X. (1999). Development prospects of China power industry. *China Power, 32,* 3–14.

Chapter 5
Renewable Energy

5.1 Introduction

Recently, there has occurred some growth of renewable energy use in China. In doing so, one difficulty has arisen from the use of these new sources of energy. China's existing electricity industry is heavily regulated by administrative planning constraints, which means that it is difficult for investors to promote the development of new types of electricity generation. Planning controls have a tendency to keep in place the existing generation mix. The intervention of local governments in direct electricity trading and the lack of a quota system for renewable energy levels have also proved problematic.

Despite these controls, some electricity industry reform is taking place and includes such things as changes to distribution and transmission tariffs, the further trading of electricity at the wholesale level and the introduction of new specific policies designed to promote the use of renewable energy. In the latter case, these include such things as the encouragement of the purchase of guaranteed renewable energy generation and green certificates.

In the years since China's Renewable Energy Law took effect on the 1 January 2006, solar and wind electricity generation have grown steadily in use, both in terms of installed capacity levels and the generation of electricity. The installed generation capacity in renewables in China rose from only 1.1 GW in 2005 to 148.6 GW in 2016 of wind generation units, and from 70 MW in 2005 to 77.4 GW in 2016 for solar units. This has meant that the proportion of China's electricity generation that is generated by renewables has risen from 0.07% in 2005 to 4.0% in 2016 for wind, and from a negligible percentage in 2006 for solar to 1% in 2016.[1] Table 5.1 and Figs. 5.1 and 5.2 provide data on the growth of electricity generation, carbon emissions and renewable energy capacity in China.

As can be seen from the data, generation growth in renewables has been strong (Fig. 5.1), and a significant proportion of generation in China is now being taken up

[1]CNREC (2015) and (China Electricity Council 2006, 2017).

M. Xiaoying and M. Abbott, *China's Electricity Industry*,
SpringerBriefs in Energy, https://doi.org/10.1007/978-3-030-53959-7_5

Table 5.1 Sources of electricity in China, 2018 (TWh, %)

Terawatt-hours	Oil	Natural gas	Coal	Nuclear energy	Hydro-electric	Renewables	Other	Total
China	10.7	223.6	4732.4	294.4	1202.4	634.2	14.0	7111.8
Total Asia-Pacific	188.0	1485.8	7290.8	553.6	1718.5	996.0	40.9	12,273.6
Total world	802.8	6182.8	10,100.5	2701.4	4193.1	2480.4	153.8	26,614.8
%								
China	0.2	3.1	66.5	4.1	16.9	8.9	0.2	100.0
Asia-Pacific	1.5	12.1	59.4	4.5	14.0	8.1	0.3	100.0
Total world	3.0	23.2	38.0	10.2	15.8	9.3	0.6	100.0

Source BP (2019)

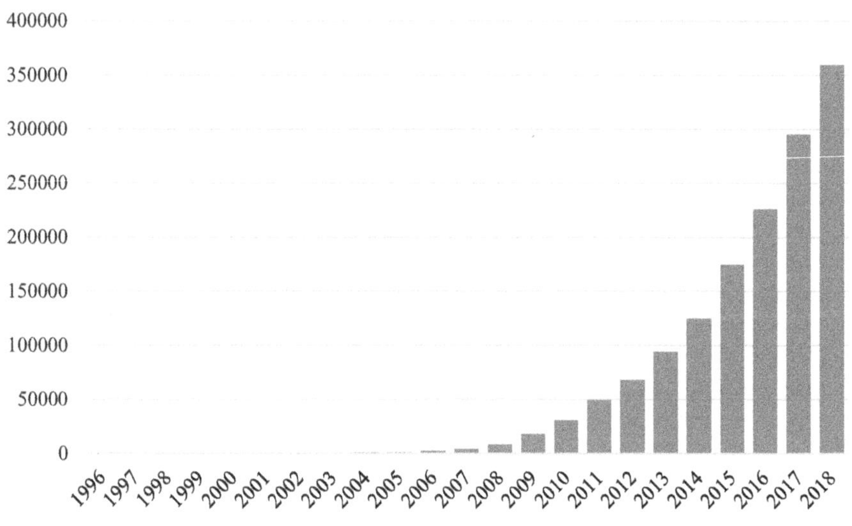

Fig. 5.1 Renewable energy capacity in China, 1996–2018 (MWs). *Source* BP (2019)

by non-hydro-renewable energy (Table 5.1). Despite this change, however, carbon emission levels are still high and rising (Fig. 5.2), which means further measures will need to be undertaken. The purpose of this chapter is to look at the growth in the use of renewable energy in China and some of the difficulties that have arisen from it.

5.2 Renewable Energy Sources

As with any commodity, electricity can be traded in markets and used in any quantity. Electricity does, however, have a number of qualities that distinguishes it from other

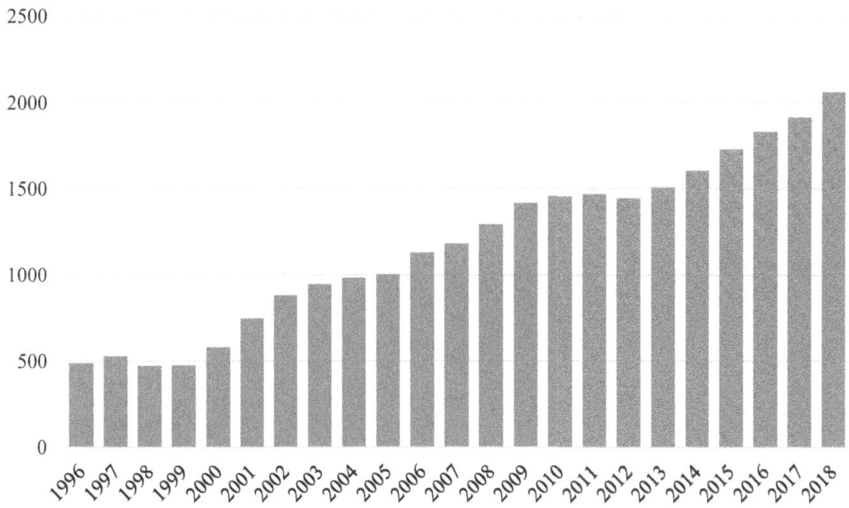

Fig. 5.2 Carbon dioxide emissions in China, 1996–2018 (million tonnes). *Source* BP (2019)

commodities. It can only be stored to a very limited degree in a commercially viable way and generally must be consumed as soon as it is generated.[2] This means that its supply and demand at any given point in time must be in balance across the electricity grid. In a wholesale electricity generation market, the commodities that are traded do not only include electricity (usually measured in megawatt-hours) but also ancillary services. Ancillary services maintain the proper flow and direction of electricity, address imbalances between supply and demand and help the system recover after a system event. Electricity can be traded in forward markets, in the form of long-term contracts, and in spot markets.

The increased use of renewable energy sources in an electricity system increases the variability and uncertainty of the supply of electricity and means that a greater level of flexibility on the part of the system is needed to respond to changes in supply and demand. There is a number of ways that this flexibility can be created, with the costs of each varying. In the past, the experience of those countries that have a high use of renewable energy (i.e. the USA, Germany and Spain), where electricity grids are operating with between 20 and 30% renewable energy, has shown that electricity market mechanisms can enhance system flexibility.[3]

Non-hydro-renewable electricity generation has characteristics that make them different from other sources of electricity. Solar and wind are both weather-dependent. Wind electricity generation can be highly variable, both in terms of the time of the day and in terms of location (see Pictures 5.1 and 5.2). Generally, wind energy is generated more at night-time when demand is low. Solar in contrast can only

[2]Large-scale battery storage is currently being developed in various parts of the world, and large water storage is a form of hydro-electricity storage.

[3]RAP (2014).

Picture 5.1 Wind generators in China

Picture 5.2 Solar panels and wind turbines in China. *Source* Authors' collection

generate electricity during daytime when the sun shines. Because of these weather dependencies, the available supply of renewable energy at any given point in time is inflexible and dependent on the changes in the weather. In the absence of flexibility in the electricity supply, or in demand-side responses, the balance of electricity supply and demand can only be maintained at the expense of grid security with frequent curtailments, or economic efficiency.

As the volume of renewable energy generation available can be only be forecast one day ahead with any reasonable degree of accuracy (through weather forecasts), renewable energy is usually traded in electricity spot markets, including day-ahead, intra-day and/or real-time markets. In electricity markets, the schedule of electricity generation is determined by the outcome of energy trading resulting from the interaction of the demand curve (the purchase amount and price offered by electricity buyers) with the supply curve (the sales amount and price offered by electricity sellers).

Another characteristic is that although renewable energy has significant fixed capital costs, it has a near-zero or zero marginal cost in its operation. In some cases where production-based subsidies exist to promote the use of renewable energy, this marginal cost can even be negative. Once established, therefore, renewable energy sources are often competitive in wholesale electricity markets and are able to be regarded as a priority dispatch. Low competitive bids from renewable energy sources, based on these low marginal costs, means that less competitive conventional electricity sources can forced to reduce their operation.

In the German case, for instance, on the 30 April 2017, 85% of electricity consumption was provided for by renewable energy, including wind, solar, biomass and hydro-electricity. This high level of use was largely attributed to the way in which the electricity spot market operated, the strong regulations that promoted renewable energy, a strong cross-border transmission network an advanced dispatching technology. During the period of the 25–30 April 2017, the highest clearing price in the day-ahead market was 0.053 euro/kWh, when solar and wind generation were very low. The lowest clearing price was 0.0075 euro/kWh which was reached at 2 pm on the 30 April.[4] This price mechanism in the spot market had the effect of encouraging other sources of electricity, such as coal, to undertake retrofitting to enhance flexibility, to generate less when electricity prices are low and generate more when electricity prices are high. What this means is that the increased use of renewable energy can create powerful incentives for the existing, conventional sources of electricity to change the manner in which they operate.

For renewable energy to operate effectively, flexibility is also required on the demand side as well as on the supply side. In a real-time market, time-of-use tariffs can influence the behaviour of consumers so as to balance electricity supply and demand. In Germany and the USA, the establishment of a real-time electricity market has encouraged the promotion of price-based, demand-side responses. As electricity prices are lowest at night-time, because of the lower demand, and as night-time is when wind generation is at its highest consumers have been encouraged to use electricity more at night through lower time-specific charges.

As mentioned earlier, the development of markets for ancillary services can help. Ancillary services refer to a broad range of functions that are required by system operators and are provided by network users (generators, customers) or system assets. These services enable system operators to maintain the stability and the integrity of distribution and transmission systems as well as the quality of service. Ancillary

---

[4] Huang and Wang (2017).

services include such things as remote automatic generation control, voltage control, spinning reserve, standing reserve, frequency control and black start capability. A high level of use of renewable energy means that there is an increased need for various ancillary services. The combination of the real-time balancing of markets and ancillary service markets helps to value these services correctly and therefore is conducive to an increased use of renewable energy.

In various countries, steps have been undertaken to develop markets for ancillary services. In Denmark, for instance, the use of renewable energy has been encouraged by making use of frequency regulation and automatic generation control to ensure the stability of the system frequency. In the real-time balancing market, the system operator dispatches the units that make the lowest offer to balance the system, thereby achieving system balance in an economically efficient manner. In Denmark as well, system stability has been enhanced by a greater interconnection of the national market with neighbouring ones. An electricity market made up of a large balancing area, with a transmission of electricity between various regions, is more likely to be able to balance the intermittent supply of electricity from renewable sources. Higher interconnection and transmission capacity also enables the optimal use of surplus generation, alleviates the problem of daily and seasonal demand peaks and reduces the need for backup generation. In Denmark, more than 40% of electricity is generated from renewable sources (mainly wind) and the existence of a market-based exchange with the neighbouring countries of Sweden and Germany is the main way in which it copes with this high level. Denmark sells electricity to these countries during periods of high wind energy generation and imports electricity at times of low wind energy generation.[5]

The use of renewable energy has faced some problems in China, and elsewhere, especially in terms of the intermittent nature of its supply. The problem of the large-scale curtailment of wind generation first took place in Inner Mongolia in 2009, and these occurrences later spread to other parts of the country in 2010. In the years 2011–2015, the average curtailment rate for wind generation in China was approximately 15%.[6] The most affected regions of the country that experienced this problem were North China, Northeast China and Northwest China, typically in Gansu, Xinjiang, Jilin and Inner Mongolia Provinces. In these major regions, in 2016, wind curtailment rates ran as high as 43, 38, and 30%, respectively.[7] As well as the curtailment of wind generation solar PV curtailment also began to be a problem in 2013. The national average curtailment rate for solar PV during the years 2013–2016 was around 15%, a level similar to that of wind. The main regions for solar PV curtailment were Northwest China, including Gansu, Xinjiang, Ningxia and Qinghai.[8]

The intermittent nature of solar and wind generation has raised concerns among government agencies in China, as well as in the renewable energy industry itself. The main factors that have been identified as being problems associated with the

[5]Ea (2015).

[6]Zhang (2016).

[7]NEA (2017a, 2017b).

[8]Greenpeace (2017).

development of renewable energy are the geographical mismatches between the location of solar and wind resources and the load centres, an inflexible generation mix which is dominated by coal-fired electricity generation, the poor nature and location of the existing transmission capacity compared to renewable energy sources, and the lack of energy storage capabilities.

An additional problem is the governance of the industry by central institutions that lack the flexibility in their operation and which have not been able to respond quickly enough to assist in the integration of renewable energy. Studies undertaken by the Regulatory Assistance Project and the International Energy Agency have raised questions about the nature of the Chinese electricity industry regarding these concerns. Cheung (2011) and Kahrl et al. (2011) have argued that the rigidity of China's electricity industry governance runs at odds with the need for flexibility in managing the intermittent nature of renewable energy. They argued that it was important that flexible electricity markets be developed so that renewable energy can operate more effectively.[9]

In addition, Kahrl and Wang (2014) have argued that China's approach to the dispatch of electricity is one of the major reasons for the frequent curtailment of renewable energy. In China, the dispatch of electricity is still carried out on the basis of plans developed at the local level and is often designed to address local economic priorities. This, they argue, undermines the use of renewable energy.[10] Pricing for both retail and wholesale electricity also remains under the control of the Chinese central government, which means that incentives to create flexible responses on the part of generators and consumers have not been developed.[11] The use of administrative mechanisms rather than economic ones for the procurement of ancillary services has also led to an undersupply of these services. Dupuy et al. (2015) and Crossley (2015) have also further argued that the lack of incentives for the grid companies to develop solutions to reducing curtailment is another reason for its occurring.[12] Although a limited level of renewable use can be managed by the present system, further reforms will need to be undertaken if a much higher share of the industry is to be taken up by renewable energy.

5.3 Chinese Electricity Markets Reform and Renewable Energy

As mentioned previously since 1985, China has gone through a number of phases of electricity market reform. The first phase of reforms in 1985 sought to encourage local government and private firms to invest in new electricity generation capacity. This policy was successful in that it led to a surge of investment in the industry

[9]Kahrl et al. (2011).

[10]Kahrl and Wang (2014).

[11]Dupuy et al. (2015), Dupuy (2016) and RAP (2008).

[12]Pollitt et al. (2017)and Dupuy and Wang (2016).

at the local level, as well as a diversifying away of central government control of generation capacity towards more localised operators. Later reforms in the late 1990s separated government functions from commercial management, which in turn led to the reforms in 2002, which moved China's electricity industry away from a being a single, vertically integrated utility, to an industry with two grid companies (the State Grid Corporation of China serving most of the country, and the China Southern Grid Corporation the five southern provinces), five large generation companies and a large number of other companies. As well, five regional grid companies were created as subsidiaries of the State Grid Corporation of China.

As noted in the previous section, the need to integrate a growing share of renewable energy with its more unstable supply has meant that there is a greater need for flexibility in the use of other generation capacity. This in turn meant that there must be a more flexible and efficient planning, scheduling and dispatch processes. The reforms to date have contributed to bringing this about, but further work is required. The governance arrangements that have developed over the years have retained some restraints, including such things as: the 'equal shares' generation dispatch approach, the government regulated electricity prices, the limited inter-provincial trade in electricity and the lack of formal compensation mechanisms for electricity ancillary services.

In terms of the dispatch of electricity in China, despite the reforms, it is still guided by instructions from planning agencies within provincial governments. Annually, the provincial governments allocate the total electricity generation hours to each generation plant based on the electricity demand forecast for the following year. Units within the same category are allocated the same annual generation hours to guarantee an equal chance for all investors and are paid a benchmark wholesale tariff set by the provincial government based on the embedded cost of a new unit. This tariff is based on the unit fixed costs divided by the operating hours, plus the unit variable costs. This dispatch system creates a conflict with the priority dispatch of renewable energy required by China's Renewable Energy Law of 2009, and as well undermines the ability of renewable energy sources to compete on the basis of lower marginal costs and greater environmental benefits.[13]

Price regulation of wholesale and retail prices by the central government also inhibits the development of renewable energy sources. The Chinese Government sets benchmark wholesale tariffs for generation plants that depend on a combination of factors such as location, technology and feedstock that broadly reflect local market conditions. This benchmark provides the basis for negotiations between individual generation plant operators and the provincial governments. The final tariff established is intended to allow the plant to make a reasonable return on its owner's investment, given that age of the plant and the number of operating hours allocated under the 'equal shares' approach. This method for setting wholesale tariffs encourages thermal plants to maximise their output, but does not encourage a degree of flexibility in dispatch. In addition, the Chinese Government also sets final retail tariffs, which can vary between different cities and provinces, and between different consumer

[13]International Energy Agency (2006).

categories. The distribution and transmission tariffs established effectively are the differences between the wholesale and retail tariffs. Price regulation inhibits the variability of the system in terms of dispatch, which in turn creates disincentives to the use of renewable energy sources.

Regarding the provision of ancillary services, the country's grid-connected electricity plant has long provided these services, but without there being any ancillary service markets. Since the reforms in 2002, ancillary services in China have been categorised into 'basic' ancillary services (basic peak regulation, primary frequency and basic reactive regulation) and 'paid' ancillary services (automatic generation control, spinning and standing reserve, paid peak regulation, paid reactive electricity regulating and black start). While basic ancillary services are mandatory and provided free of charge by generator companies, other ancillary service providers receive payments. Electricity generation units are currently obliged to provide these basic ancillary services arranged by the dispatch agencies based on the principle of 'dispatch whenever there is a need'. There is no standard mechanism for procuring the compensated ancillary services, and the payments to grid-connected electricity plants which provide ancillary services are usually less than 0.30% of generation revenue, barely covering the costs of the ancillary services. The lack of a market for ancillary services means that there has been an underinvestment in provision of them, which in turn inhibits the flexibility of the system to adapt to a variability of supply. This makes it difficult for system ability to cope with the greater supply instability caused by the growing use of renewable energy.

One way to help provide greater supply stability, both with and without the greater use of renewable energy, is the greater use of inter-regional trade in electricity. Despite the growing degree of interconnection between regions in China, the volume of inter-regional trading of electricity remains low. In addition, some local governments have intervened in inter-provincial electricity trading to limit its development. In 2014, for instance, the Jiangxi Energy Administration Bureau required that the annual electricity imported by the province should not exceed 10 TWh, and electricity imported temporarily (within three days) and in short term (3–10 days) should not exceed 1 TWh. This sort of provincial development is important as it can be used either to assist or prevent the development of Chinese renewable energy.

In China, provincial governments are still directly involved in investment and planning decisions, an interest they have retained because of the role the industry plays in providing employment and revenue to the governments. The provincial governments control or guide the siting of plants, the financing of construction through ownership of local banks, the annual allocation of generation hours and through the setting of wholesale and retail tariffs (although guided by central government rules). Provincial governments also directly invest in the construction of new generation plant on their own account. As well the creation of greater balancing grid agencies after the 2002 reform, within the State Grid Corporation, through the use of five regional grid companies was undermined when the assets of these companies were allocated to provincial subsidiaries. These five regional grid companies effectively then lost their autonomy and so were made the subject of the equal dispatch approach.[14]

A growing recognition that the reforms of 2002 were insufficient to create an efficient electricity industry led in March 2015, to the State Council and the Central Committee of the Communist Party releasing 'Several Opinions on Deepening Power Sector Reform Document#9'. Following the issuance of Document#9, six accompanying documents were released, detailing the specific reform measures. The overall objective of the new reforms was to create a more market-based electricity industry.[15]

The new reforms follow international trends in that they involve regulation of the distribution and transmission parts of the industry and the introduction of competition into the retail and generation parts of the industry. The grid companies were to be stripped of their monopolies in the retailing of electricity, allowing other firms, including many privately owned ones, to engage in electricity retailing.

According to the reform documents, the path towards an electricity market involves the:

- Phasing out electricity generation and consumption planning, and the deregulation of electricity tariffs in the potentially competitive parts of the industry (generation and retailing);
- Expansion of the scale and scope of participation in the direct trading of electricity;
- Gradual establishment of market-oriented mechanisms for the inter-provincial trading of electricity; and
- Identification of qualified pilot areas to build more complete electricity markets, including long- and mid-term markets as well as spot markets, and to later expand the scope of these pilot areas.[16]

In contrast to the reforms of 2002, the new ones have placed more emphasis on environmental and emissions issues. In terms of renewable energy, this means that:

- Renewable energy is to be given priority in generation planning and dispatch;
- The proportion of renewable energy in inter-provincial and inter-regional electricity trading is to be increased;
- Ancillary services markets and cost-sharing mechanisms are to be created; and
- Electricity plants are to be encouraged to engage in the expanded delivery of electricity ancillary services.

In the period since 2015, four types of reforms have been piloted in China, including changes to distribution and transmission tariffs, retail reform in a number of provinces and municipalities and reforms in new distribution business in a range of pilot areas. Furthermore, electricity trading organisations have been established and are now operating either at the national, regional or provincial levels.

The latest reforms of pricing and electricity generation in China have three parts: distribution and transmission tariff reforms, the promotion of the direct trading of

---

[14]Dupuy and Wang (2016).

[15]CCCP and the State Council (2015).

[16]Zhao (2004).

electricity and the reduction in the importance of central planning. Distribution and transmission tariff reforms change the ways in which the grid companies generate revenue and profits. In the distribution and transmission parts of the industry, tariffs should be set by a regulator to promote economic efficiency by being based on a revenue cap that provides for allowable costs plus a reasonable profit.[17] A revenue cap regulatory system based on this approach breaks the link of distribution and transmission tariffs between wholesale and retail tariffs, and helps to initiate the deregulation of generation and retail. This it does by allowing for competition between generators, further inter-regional trading, long-term and mid-term trading as well as spot trading. For renewable energy use, it means that the advantage of low marginal costs can be taken advantage of, which had previously not been the case.

The direct trading of electricity allows electricity users to bypass the grid company and to negotiate prices directly with generators. This means that prices of electricity are determined more by market forces instead of by government regulation. The direct trading of electricity was introduced as part of the reforms from 2002, however, as some provincial governments used this system to give preferential tariffs to favoured, established generators, renewable energy sources were often prevented from operating.

The reforms introduced in 2015 emphasised the need for the promotion of the existing provincial-level pilot programmes in renewable energy, and since 2015, the direct trading of electricity has occurred in the northwest and northeast of China between renewable energy generators and large-scale consumers. In addition, the direct trading of generation rights has taken place between renewable energy generators in Northwest China and thermal generators in places such as Henan, Shanghai and Chongqing.

The reduction of planning controls on the operation of electricity generation since 2015 has taken place through the introduction of a prioritised dispatch system that makes use of renewable energy during peak periods. The on-grid tariff for these sources can either be set by government or established in the wholesale market. Local governments are increasingly being required to reduce their generation quotas for the existing coal-fired electricity plants. If, in the future, these measures become fully implemented, traditional electricity sources will no longer take priority of dispatch.

As well as the introduction of a freer wholesale market for electricity there has also been introduced more specific measures to promote the use of renewable energy. These include the requirement for the full purchase of guaranteed renewable energy generation and the creation of a green electricity certificate system. The Renewable Energy Law, enacted in 2006 and amended in 2009, requires a full purchase of all renewable energy generation. Document#9 and several supporting policies have reaffirmed this principle. These measures were limited in impact, and on 28 March 2016, the Chinese Government promulgated the Management Measures for the Full Purchase of Guaranteed Renewable Energy Generation (the Measures).[18] This document introduced a mandated guarantee that grid companies purchase output

[17] Dong (2017).

[18] National Development and Reform Commission (2016).

from renewable generators in regions of the country that have been experiencing reduced use, at least up to an allocated number of hours. The guaranteed number of hours is based on an assured internal rate of return for renewable energy generators. It also introduced direct compensation for renewable energy generators for reduced use. In the case where this results from non-renewable generators infringing on absorption space and transmission capacity, then the non-renewable generators are responsible for paying compensation. Alternatively, if the reduced use is due to the grid company's misconduct, then the grid company must take responsibility for compensation. As well the National Development and Reform Commission and the NEA were made responsible for planning the allocation of the guaranteed number of hours for each type of renewable generation. The measures also meant that beyond the allocated hours which are guaranteed purchase, renewable energy generators are permitted to negotiate contracts with consumers and these contracts are given priority dispatch rights. Renewable energy generators are encouraged to participate in mid- to long-term trading and spot market competition. The seller of this traded electricity receives the negotiated price, as well as the national subsidy, which is the difference between local feed-in tariffs for renewable energy and the local benchmark tariffs for de-sulphurised coal-fired electricity.[19]

Altogether, these measures reduced the power of provincial governments and meant they should fully reserve for wind and solar electricity the guaranteed generation hours required by the central government, when designing and implementing electricity generation and electricity trading schemes. Electricity grid companies are to enter into contracts with wind and solar electricity generators for annual priority generation purchase of guaranteed renewable energy.

A further measure was taken on the 6 February 2017, when the Notification on the Experimental System of Verification, Issuance, Voluntary Purchase and Transaction of Green Electricity Certificate of Renewable Energy was issued jointly by the NEA, the National Development and Reform Commission and the Ministry of Finance. The green certificate under this measure is an electronic certificate, which bears a unique government-issued identification code which is given to generators of non-hydro, grid-connected renewable energy. Each certificate is equivalent to 1 MWh of electricity and has first been run as a trial. Certificates have also been issued to onshore, grid-connected wind and solar PV electricity generators, with other renewable sources, including distributed solar PV electricity being excluded. In order to apply for the green certificates, wind and solar PV electricity generators need to be able to submit verification documents to the Information Management Centre of the NEA. Once verified, the Centre then issues the certificates monthly.

The pilot program of voluntary purchase of these certificates began on the 1 July 2017, with the prices of certificates being determined by negotiation between buyers and sellers. The price is not below that of the renewable energy subsidy level received for the same quantity of electricity. These certificates are not transferable, and purchasers are not eligible for other renewable energy government subsidies. After the completion of the voluntary scheme, a compulsory scheme was launched

[19]Qin (2016)and National Development and Reform Commission (2017).

in 2018.[20] The green certificate system means that solar and wind generators receive additional revenue in a more timely fashion than was possible via government subsidy payments. It does, however, also have the effect of raising overall electricity prices to final consumers.

The direct trading of electricity has been one of the main parts of the most recent reforms. In practice, however, some provinces have forced renewable energy generators to lower their prices, in order to prevent retail prices rising too far. In some cases, renewable energy generators have expressed the view that the current market trading mechanism has been used more to reduce tariffs for some energy-intensive electricity consumers. So far, most of the direct electricity trading has been within provinces, and regional electricity markets are still underdeveloped. Intra-provincial electricity trading generates more taxes for provincial governments than inter-provincial electricity trading and is therefore generally favoured by provincial governments. This occurs because taxes on electricity generators and electricity consumers go to provincial governments, while taxes on grid companies go to the central government. In addition, as, in recent years, supply has tended to exceed demand, most provinces have little incentive to import electricity from other provinces. Finally, the high cost of transmission and line losses for inter-provincial electricity flows are still an impediment to inter-provincial trading. Although the reforms in theory should promote trading and the greater use of renewable energy, there is still some opposition in practice to them from the provincial governments and the grid companies. As the electricity markets in China are not as yet properly developed, the effective implementation of the green certificate system is problematic. The recent reforms do require a degree of cooperation from provincial governments in implementing the reforms, and however, they often seek to promote industrial development by keeping retail tariffs low, and this conflicts with the need to raise prices to encourage investment in renewable energy.

The development of electricity spot markets can help to provide more transparent operations and information to market participants and can become the basis for the determination of long-term contract prices as well as prices for ancillary services. The operation of an electricity spot market, however, is complex and potentially quite costly. The long-term benefits can be significant, however, and as they can cover multiple regions, they can help to promote the construction of a greater level of interconnection between regions.

Greater dispatch between regions would reduce generation costs and as well reduce overall emissions by allowing wind and solar generation to be used, located in the regions where they operate best. As well it would help to reduce the variability of both load and generation and furthermore reduce the level of market power that some generator companies possess in some provincial markets.[21] Mechanisms for the setting of regional distribution and transmission tariffs in a more rational way need to be developed along with the removal of any barriers to inter-regional and

---

[20] National Development and Reform Commission (2017).

[21] Wang (2017).

inter-provincial trade. Incentive mechanisms also need to encourage grid companies to play a greater role in the construction of regional markets.

5.4 Conclusion

Although there has been a dramatic growth in the use of renewable energy in China in recent years, there have been some difficulties with this change occurring. The nature of the existing electricity industry governance arrangements still involves the heavy regulation of participants as well as planning by central authorities. These authorities often have shown a bias towards the existing coal-fired plant and the expense of new entrants. Recent reforms have aimed to promote the increased use of renewable energy. These reforms include changes to distribution and transmission tariffs setting, more direct trading of electricity, the opening up of electricity generation construction, the establishment of new ancillary services purchase mechanisms, the full purchase of guaranteed renewable energy generation and the introduction of a green certificate scheme.

Looking forward, however, there are still some difficulties that will need to be overcome if growth of the use of renewable sources of energy is to occur. First of all, the direct trading of electricity is affected by the low tariffs set by local governments. This in turn has an impact on the use of renewable energy sources. Local economic priorities often conflict with the national goal of promoting renewable energy. The implementation of the green certificate system in the absence of fully competitive electricity markets also is proving problematic, and measures will need to be undertaken to correct this.

References

BP. (2019). *BP statistical review of worked energy.* London: BP.

CCCP and the State Council. (2015). Several opinions on deepening the power sector reform [Document#9] (in Chinese). Beijing: CCCP and the State Council. http://www.net21.com/news/show-64821.htm.

Cheung, K. (2011). *Integration of renewables: Status and challenges in China.* Working Paper, Paris: IEA.

China Electricity Council. (2006). *Bulletin of the national power industry statistics in* 2005. http://suo.im/2xqST.

China Electricity Council. (2017). *Bulletin of the National Power Industry Statistics in 2016.* Beijing: CEC. http://suo.im/2Fdwqg.

CNREC. (2015). *China renewable energy industry development report in 2015.* Beijing: CNREC.

Dupuy, M. (2016), *China power sector reform: Key issues for the world's largest power sector.* Beijing: RAP. http://suo.im/Fipdg.

Dupuy, M., & Wang. X. (2016, April 8). *China's string of new policies addressing renewable energy curtailment: An update* (in Chinese). http://suo.im/2eJawp.

Dupuy, M., Crossley, D., Kahrl, F., & Porter, K. (2015). *Low-carbon power sector regulations: Options for China*. Beijing: RAP.
Dong, Y. (2017). *Ten years' retrospect of direct electricity trading at the period of new electricity reform* (in Chinese). Beijing: March. http://suo.im/3I5rB0.
Ea (2015). *The Danish experience with integrating variable renewable energy*, Study on behalf of Agora Energiewende. http://suo.im/1006RQ.
Greenpeace (2017). Research on the coordinated benefits of wind and solar power. http://suo.im/1lqTII.
Huang, Y., & Wang, Y. (2017). The Germany's experience with new energy integration and implications for China (in Chinese). *Journal of State Grid*. http://suo.im/1snwxa.
International Energy Agency. (2006). *China's power sector reforms: Where to next?*. Paris: IEA.
Kahrl, F., & Wang, X. (2014). *Integrating renewables into power systems in China: A technical primer—power system operations*. Beijing, China. http://suo.im/3I5v1h.
Kahrl, F. & Wang, X. (2015). *Integrating renewable energy into power systems in China: A technical premier–electricity planning*. Beijing: RAP. http://suo.im/2WcipP.
Kahrl, F., Williams, J., Juanhua, Ding, & Junfeng, Hu. (2011). Challenges to China's transition to a low carbon electricity system. *Energy Policy, 39*, 4032–4041.
National Development and Reform Commission (2016, March 24). *Management measures for full purchase of guaranteed generation* (Fagai Nengyuan [2016]625]) (in Chinese). Beijing: NDRC. http://suo.im/48PyrH.
National Development and Reform Commission (2017). *Notification on the experimental system of verification, issuance, voluntary purchase and transaction of green electricity certificate of renewable energy* (Fagai Nengyuan#[2017]132) (in Chinese). Beijing: NDRC. http://suo.im/2iJyOP.
NDRC and NEA (2017, March 29). *Notification on the orderly release electricity generation and consumption planning* (Fagai Yunxing#[2017]294) (in Chinese), Beijing: NDRC and NEA. http://dwz.cn/5J19Bk.
NEA. (2015). *Officials from NEA answered correspondents' questions on electric power sector reforms* (in Chinese), Beijing: NEA. http://suo.im/MWZx.
NEA. (2017, January). *The status of wind power integration in 2016*, Beijing: NEA. http://suo.im/4dw21T.
NEA. (2017, April 10). *Circulation on monitoring and evaluation of national renewable energy development in 2016* (Guoneng Xinneng#[2017]97) (in Chinese). Beijing: NEA. http://suo.im/hcXdq.
Pollitt, M. G., Yang, C.–H., & Chen, H. (2017). In: *Reforming the Chinese supply sector: Lessons from international experience*. Cambridge Working Paper, Cambridge: Cambridge University. http://suo.im/.
Qin, H. (2016). *An in-depth interpretation of full purchase of guaranteed renewable energy* (in Chinese). http://www.wusuobuneng.com/archives/29362.
RAP. (2008). *China's power sector: A backgrounder for international regulators and policy advisors*, Beijing: RAP. http://suo.im/3b7JXW.
RAP (2014). *Low-carbon power sector regulation: International experience from Brazil, Europe, and the United States*, Beijing: RAP. http://www.raponline.org/document/download/id/7432.
Wang, P. (2017, July 10). Break the inter-provincial barriers make joint efforts by government and enterprises to push forward regional electricity market, (in Chinese). Beijing: China Energy Newspaper. http://suo.im/1CUoNR.
Zhang, Y. (2016, February). Wind curtailment statistics in China: 2011–2015, wind energy (in Chinese). http://suo.im/ixAKe.
Zhao, J. (2004). Reflection on the reform of China's electricity industry. *Energy of China, 26*, 8–12.

Chapter 6
The Future

6.1 Introduction

Given the past growth of the industry and the overriding imperative to create new generation capacity in the past, it is important to try and see how growth in general will shape the industry into the future. The purpose of this chapter, therefore, is to describe and identify some of the major components and variables that must be incorporated into any forecast of future Chinese electricity demand and use these to make some inferences of the future. To do this, the construction of a forecasting model is necessary that is then used to forecast growth in electricity demand in China. The prospects are that it might not only be used to estimate future demand for electricity in countries like China, the primary focus, but also could be adapted to be used for other East and South-East Asian countries as well.

One focus of the model is on determining the future prospects for the growth of renewable generation capacity in Asian markets. In providing this description an account of both the general background to the electricity industry will be given, along with an account of the model itself. In undertaking this work, it will be possible to provide greater certainty to investors in electricity generation capacity (renewable and non-renewable) and therefore assist the financing for innovation, entrepreneurship and renewable energy development in developing countries. The model itself is developed for three classes of customers (residential, commercial and industrial) and for two time periods (short and medium term), which makes it unique among present forecasting models in China.

The first requirement, therefore, is to identify the key variables that would go into the development of a forecasting model of demand in the Chinese electricity market. The identification of these variables is a key in being able to develop a forecasting model that will achieve a degree of accuracy that will be useful to both policy makers and investors in the Chinese electricity industry. The second requirement involves using the model to show what electricity demand in China will look like over the medium to long term.

M. Xiaoying and M. Abbott, *China's Electricity Industry*,
SpringerBriefs in Energy, https://doi.org/10.1007/978-3-030-53959-7_6

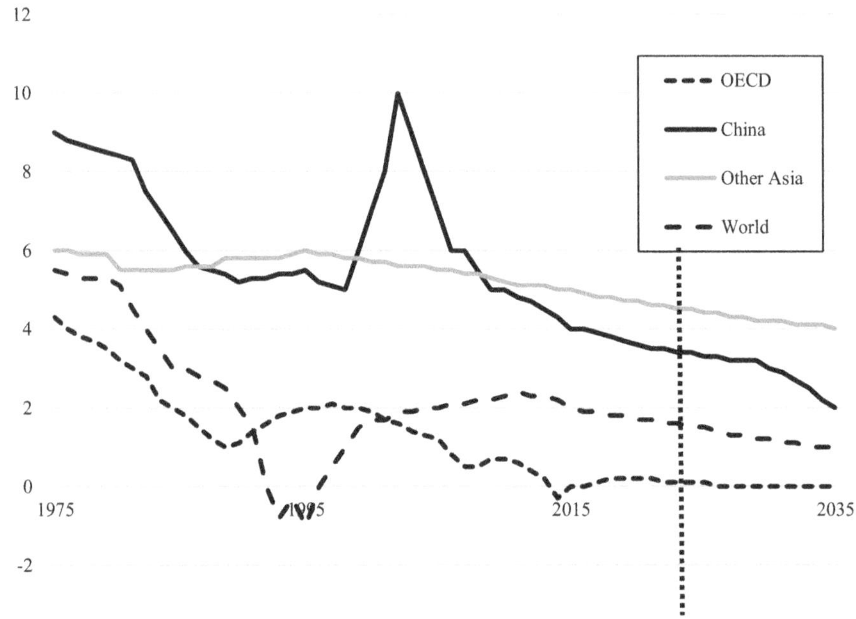

Fig. 6.1 World energy consumption growth by region; ten-year average, actual and expected (%, per annum). *Source* BP (2019)

In the first part of the chapter, a general background to the nature of the electricity industry and the state of renewables in China is provided.[1] In the following section, some of the general principles of demand estimation in the electricity industry are provided. This is followed by a section on the model being developed and the data that is being collected. In the next section, some conclusions are made about the nature of the industry in the future.

6.2 Background

China is the world's leading country in electricity production not only from conventional sources of electricity but also from the view of renewable energy sources. Production and consumption of electricity in China is still growing at a steady rate although this growth is currently slowing down (see Fig. 6.1 for growth from 1975 onwards compared to other regions of the world). Despite this slowdown, the average energy consumption in China per capita is still significantly below that of other countries and so the demand for energy is expected to grow for several years to come and China's share of world energy consumption is expected to grow through to 2035.

[1] ABARE/Asia-Pacific Economic Cooperation Energy Working Group (2002), Blackman and Wu (1999) and Chow (1993).

In terms of the production mix, China's renewable energy sector is growing faster than the use of fossil fuels and nuclear power capacity; although the bulk of this sector is concentrated in hydro-electricity, with wind and solar power both growing in importance. Including hydro-electricity, renewables make up about 23% of electricity generation, with most of the remainder provided by traditional coal power facilities. This level is expected to grow in the future.

In terms of its sheer size China is today, therefore, the world's largest generator and consumer of electricity[2] and electricity generation has more than doubled in size since 2005, although generation, which is mostly driven by economic and industrial demand, decelerated after the global financial recession in 2008 mainly a result of significant slowdown of activity in heavy industries, especially the steel industry.

Coal and hydro-electricity are the leading sources of the country's electricity generation and installed capacity; however, the country is moving to generate more electricity from nuclear, renewable sources and natural gas in efforts to address environmental concerns and to diversify its electricity generation fuel sources. Installed capacity is expected to grow over the next decade to meet rising demand, particularly, in large urban areas in the eastern and southern regions of the country, and to diversify in terms of fuel source, meaning that accurate estimates of demand growth are needed to ensure adequate investment occurs.

Of primary energy consumption in China around two-thirds is from coal and another 20% from oil and 7% from natural gas (see Fig. 1.1 for 2018). Hydro-electricity makes up 8%, the main form of renewable energy. The proportion from coal is high by international standards, as is the contribution made by. This is offset by a lower use of natural gas in China compared to many countries. The situation is also true in the case of electricity generation.

The proportion taken by coal is falling and is well below the share it held twenty or thirty years ago and is expected to decline in the years up to 2040. Most other forms of energy source are growing in importance. This trend can also be seen in the fall in the ratio of coal demand to GDP growth, which has fallen considerably since 2005 and is expected to fall further in the future. Coal-fired electricity production is still undertaken on a massive scale in China. As coal-fired stations tend to create the most carbon emissions of the various types of electricity generation the Chinese Government is keen to expand production of electricity from sources such as hydro, natural gas, nuclear and non-hydro-renewables.

In addition to concerns over emissions, the government of China also sees renewables as a source of energy security. China's Action Plan for the Prevention and Control of Air Pollution issued by China's State Council in September 2013, for instance, provides an illustration of this attitude of the Chinese Government's desire to increase the share of renewables in the country's energy mix. In this plan, renewables are favoured because unlike oil, coal and gas, the supplies of which are finite and increasingly imported, and therefore the subject of geopolitical tensions, renewable energy systems can be built and used wherever there is sufficient water, wind and sun. In terms of the actual renewables themselves, as of 2020 hydro-electricity remains by

[2]United States, Department of Energy, Energy Information Agency (2020).

far the largest component of renewable electricity production. Wind power provides the next largest share followed by bio-fuels. Solar PV power started from a low base and has grown rapidly since then.

Of the non-hydro-renewables, China has become the world's largest producer of photovoltaic power and is the world's largest maker of wind turbines. China has become a world leader in the manufacture of solar photovoltaic technology, with its six biggest solar companies having a combined value of over $15 billion. In addition, China has the largest wind resources in the world. It is the stated aim of the Chinese Government to have 210 GW of wind power capacity in 2020. In addition to wind and solar, China has become the world's third-largest producer of ethanol-based bio-fuels (after the USA and Brazil) and at present ethanol accounts for 20% of total motor-car fuel consumption in China. Bioenergy is also used at the domestic level in China, both in biomass stoves and by producing biogas from animal manure. Geothermal resources in China are also abundant and widely distributed throughout the country. Scope for expansion of this energy source, however, is constrained. Despite these developments, China still produces most of its energy from coal and emits more carbon dioxide than the next two largest emitters combined (the USA and India).

In terms of the Chinese Government's formulation of policy towards the electricity industry, after the dissolution of the Energy and Industry Department in 1993, the industry in China has been supervised by multiple organisations such as the National Development and Reform Commission, the Ministry of Commerce and the State Electricity Regulatory Commission. In 2008, the National Energy Administration was founded, and in January 2010, the State Council decided to set up a National Energy Commission. The commission is responsible for drafting a national energy development plan, reviewing energy security and major energy issues and coordinating domestic energy development and international cooperation.

The commission in the time that it has operated has provided detailed incentive policies and programmes include the Golden Sun programme, which provides financial subsidies, technology support and market incentives to facilitate the development of the solar power industry; the 'Suggestions on Promoting Wind Electricity Industry', which offers preferential policies for wind power development; and many other policies.

The NEA launched the first domestic real-time spot electricity trading markets in eight Chinese regions (Inner Mongolia, Zhejiang, Shanxi, Shandong, Fujian, Sichuan, Guangdong and Gansu) in 2018 after these regions launched power markets in 2017 for both monthly and quarterly prices.

Other programmes are still in the process of being developed. In November 2011, China approved pilot tests of carbon trading in seven provinces and cities. The pilots were intended to test the possibility of a wider scheme. Seven carbon emissions trading pilot programmes were introduced in 2013, and others followed. The national market is still planned to begin in 2020 and when the market is launched, it will be the largest carbon market in the world, given that the country is the largest creator of carbon emissions. The Chinese Government Ministry of Finance had also proposed to introduce a carbon tax, based on carbon dioxide output from hydrocarbon fuel sources such as oil and coal. Both of these measures have the potential to increase demand for renewable energy sources.

Given the growth in production and consumption of energy in China and the change in the energy mix away from coal to other sources such as hydro, natural; gas, nuclear and non-hydro-renewables, it is important for investors to have some knowledge of future demand growth. Most demand projections in China are crudely based on extrapolating present trends forward rather than on any more sophisticated approach that incorporates all important variables. The present project, therefore, takes the electricity industry as the country's most important part of the energy sector and works on developing a model of its future growth.

Rapid growth in electricity demand this past decade spurred significant investment in new power stations, particularly, in fossil fuel-fired capacity. Although much of the new investment over the past several years was earmarked to alleviate power supply shortages, the economic crisis of 2008 and the deceleration of Chinese economic growth after 2012 resulted in a slower demand growth for electricity. This volatility of demand growth highlights the great need for more accurate estimations of demand growth so that sufficient investment in capacity can be created.

6.3 Energy Demand

In order to enable sufficient and appropriate investment in new capacity, an accurate forecast is needed. Electricity demand forecasting plays a significant role in electricity system planning, energy policy designing and social stabilisation.[3] Underestimation of demand growth generally results in insufficient capacity being created and unmet demand. By contrast, overestimation of demand can lead to the creation of stranded assets and a misallocation of capital resources.[4]

Compared with developed countries, where economic and electricity systems are mature and stable, developing countries generally face more volatile demand growth, affected by social and economic conditions, unexpected events and technology breakthrough. All of these uncertainties add difficulties in conducting an accurate electricity demand forecast. In terms of the Chinese economy, there has been a growing body of literature on the means by which forecasts of Chinese electricity demand can be undertaken.[5]

To date, the empirical methods used in electricity demand forecasting include neural network, time-series analysis and multivariate analysis. Among these methods, multivariate analysis enjoys a great popularity with the advantages of simplicity and low data requirements. Although to date there have been attempts to forecast Chinese industrial and residential demand for investment decisions on capacity creation total

[3]Dergiades and Tsoulfidis (2008), Liao et al. (2016, 2017) and Son and Kim (2016).

[4]Sadeghi Keyno et al. (2009), Liao et al. (2017) and Son and Kim (2016).

[5]Adams and Shachmurove (2008), von Hirschhausen and Andres (2000) and Steenhof and Fulton (2007).

demand should be included. Furthermore, both provincial and national data can be used encompassing more than 30 provinces and municipalities,

Modelling demand is an important part of investing in the electricity industry. Energy projects tend to be capital intensive and often require long lead times.[6] For example, a thermal power station may need three to four years to build, a nuclear or hydro-electricity station requiring typically seven to ten years, if not more, and a refinery project can easily take two to four years. Given the long gestation period of energy investments and diversity of technologies, as well as economic conditions of countries and consequent constraints on the analytical choices, medium to long-term analysis is essential for energy system-related decisions and financial planning.

At the other end of the spectrum, there is also a need for energy forecasts for day-to-day operation and management of energy systems. How much electricity, for instance, needs to be generated next hour or tomorrow? This information forms the basis for unit commitment exercise (i.e. to find out which plants should be used to produce the required electricity so that the operating cost is minimised). Similarly, forecasts for six months to one year are required for business planning purposes, for regulatory approvals, to assess the prospects of the business in the coming year, etc. Thus, short-term forecasting is also important in addition to medium to long-term analysis and forecasting.

In terms of modelling of both long them and short-term demand, it is important to recognise that energy is not consumed for its own sake, but is instead used for satisfying some need and is done using appliances and equipment. Any commercial energy requires monetary exchanges and the decision to switch to commercial energies can be considered.

- First, a household has to decide whether to switch energy sources or not (i.e. switching decision).
- Second, it decides about the types of appliances to be used (i.e. appliance selection decision).
- Third, a consumption decision is made by deciding the usage pattern of each appliance (i.e. consumption decision).

All of these factors need to be captured in the modelling approach. Historically, there are a range of different approaches to capturing these affects.

6.4 Energy Demand Forecasting

The analysis of the historical evolution of energy demand and its interpretation is an essential part of energy demand analysis. Such an analysis allows for an identification of the underlying factors that affect energy demand. Various analytical methods have been used to analyse energy demand. Overall there are three approaches to forecasting energy demand which are presented below:

[6] Gianna (2011) and Bhattachayya (2011).

6.4.1 Simple Descriptive Analysis

The first approach that has been used is to use some simple descriptive indicators, three commonly used indicators are: growth rates, demand elasticities and energy intensities.

- With growth rates, generally, a trend is detected by looking at annual growth rates and then extrapolating them along a trend. In the case of China's electricity demand, this has been a common approach.
- In the case of demand elasticities, these elasticities are used to measure how much (in %) the demand would change if the determining variable changes by 1%. In any economic analysis, three major variables are considered for elasticities: output or economic activity (GDP), price and income.
- In the case of energy intensities (also called energy output ratios), these measure the energy requirement per unit of a driving economic variable.

The trend analysis then extrapolates the past growth trends and is normally done by fitting some form of time trend to past behaviour. The analysis, however:

(a) Assumes that there will be little change in the growth pattern or in the determinants of demand such as incomes, prices, consumer tastes, etc.;
(b) Finds the best trend line that fits the data. This is usually estimated by a least square fit of past consumption data or by some similar statistical methodology; and
(c) The fitted trend is then used to forecast the future. Frequently, ad hoc adjustments are made to account for substantial changes in expected future demands due to specific reasons.

Direct surveys can also be used to generate primary information essentially for the short-term, but surveys can also be used as a direct and reliable tool for demand analysis and forecasting. Such surveys ask major energy users to reveal their present consumption and future consumption plans. Through this, surveys try to account for changes in the energy-consuming sectors themselves that would affect the demand. In addition, by analysing the investment plans and programmes, changes in the supply and demand are captured.

6.4.2 Factor (or Decomposition) Analysis

For a better understanding of energy use and future energy requirements, it is important to understand the causal factors. A particular method, known as decomposition method, has been widely used. Traditionally, these methods try to identify changes in energy demand arising from several factors, the commonly used ones are: changes in economic activity (the activity effect), changes in technological efficiency of energy use at the sector level (the intensity effect) and changes in the economic structure (the structural effect). They then are used to construct trend data into the future.

6.4.3 Advanced or Sophisticated (Econometric) Techniques

The final approach is to use sophisticated demand forecasting techniques which rely on more advanced methodologies. Such techniques can be classified using alternative criteria: for example, a common method of classification is the top-down and bottom-up models. Top-down models tend to focus on an aggregated level of analysis while the bottom-up models identify the homogeneous activities or end-uses for which demand is forecast. Econometric models are grounded in the economic theories and try to validate the economic rules empirically. Engineering–economy models (or end-use models), on the other hand, attempt to establish accounting coherence using detailed engineering representation of the energy.

The purpose of the present work is to move away from the simple descriptive approach used in the past, but instead to make use of a developed econometric model incorporating several variables that impact on demand. In doing so, the intention is to concentrate on demand for electricity as opposed to other forms of energy. In doing so, however, data would need to be acquired for all types of energy in China. The modelling follows the techniques developed by Australian economists during the 1980s and 1990s and in line with conventional approaches internationally.[7] In addition, it can be used with data at a provincial level across China. It does include all three major sectors that demand electricity rather than focuses on a single sector.[8]

1. Data Required—Econometric modelling (electricity)

Two models could be constructed for electricity (short and medium term).

To do this the following data would be required.

Electricity production and consumption (broken down into classes of customers; residential, commercial and industrial).

(Sources of data include State Electricity Regulatory Commission of China; BP; International Energy Agency; FGE *China Oil and Gas Monthly*).

This is the dependent variable of the modelling. The following would be independent variables.

Population and population growth (both national and provincial).

(Sources include the United Nation's forecasts, National Bureau of Statistics, monthly and provincial data; The World Bank.)

Per capita income (national and provincial).

(Sources include the United Nation's, National Bureau of Statistics, monthly and provincial data; The World Bank; Asia Bank.)

Industry structures (both national and provincial).

(Sources include the National Bureau of Statistics, monthly and provincial data; The World Bank; Asia Bank.)

Survey data on the behaviour of residential, commercial and industrial consumers.

(Source: surveys of groups of 300 consumers undertaken periodically).

[7]Cox (1987), Australian Gas Association/ABARE (1996), Australia, Bureau of Resources and Energy Economics (2014), Bhattachayya (2011), Gianna (2011).

[8]See for instance Liao et al. (2017).

Price data on electricity and that of substitutes (coal and natural gas).

(Sources of data include State Electricity Regulatory Commission of China; BP; International Energy Agency; FGE *China Oil and Gas Monthly also the National Bureau of Statistics.*)

For short-term forecasting weather patterns.

(Data sources: the China Meteorological Administration).

Once the strength of the relations between the various variables has been determined it is possible to use the more general Chinese Government and business forecasts to generate electricity demand forecasts. In the case of the short-term movements in demand weather forecasts are used along with the other variables to generate probability estimates of demand based on temperature changes, and the changes in demand for electricity for heating and cooling purposes.

Short-run demand can be determined by the following:

$$Q = (1 + G) \times (a + b_1 T + b_2 P + b_3 W)$$

Q Daily demand
G Growth rate
T Forecast temperature
P Prior day temperature
W Forecast wind speed.

In the process, cross-elasticities are estimated which assists in constructing the model. The share of a particular fuel in total expenditure on energy is a function of its price, together with that of the prices of alternative fuels. For each fuel or energy type I, the share equation is given by:

$$S_i = \alpha_i + \beta_{ij} o P_j + \beta_{iT} T$$

where $S_i = p_i Q_i / \Sigma P_j Q_j$ the expenditure share of energy type I;

Q is the quantity of each energy type consumed
S is the expenditure share of energy type I
P is the price of energy type I
T is the time to allow for technological change in energy applications
i, j are subscripts indicating energy type (gas, electricity and other fuels).

Longer terms demand forecasts are then generated.

6.5 Forecasts

Using the above-presented approach, it is possible to arrive at some estimates of what the Chinese electricity will look like over the longer run. In terms of energy consumption forecasts, the petroleum producing company, BP, produces regular forecasts for China and other countries. These are summarised in Fig. 6.1, along with actual growth in consumption of energy for the OECD, the world as a whole and other Asian countries. As can be seen from the figure consumption growth in China for energy has been high coming up to 2020, above the rest of the world. The expectation of BP is that this growth will decline, as is has done since around 2008, and converge towards that of the rest of the world, without reaching it, by 2035.

Figure 6.2 shows summary data on the annual average growth of population, GDP and electricity generation in the past (1996–2020), and in the future (2021–2030 and 2032–204). The population and GDP growth estimates are based on those produced by the United Nations. The electricity generation forecasts are based on use of the model presented above. Population growth is already very small and expected to reach zero sometimes in the next few years. The growth the Chinese economy is also expected to decline, although still be significant. As the economy tends to shift more to an emphasis on services and away from heavy industry electricity demand growth will slow down, without ever stabilising.

In terms of the breakdown in demand for electricity, it is possible to take the forecasts produced by the United Nations and use them to show how this will change over time. Figures 6.3 and 6.4 show the past growth in demand for electricity of the various sectors (industry/transport, residential and commercial) for the past (1996–2000) and the future (2021–2030 and 2031–2040). Growth in demand in each sector will growth, but at a diminishing rate as time passes. A change will also occur in the

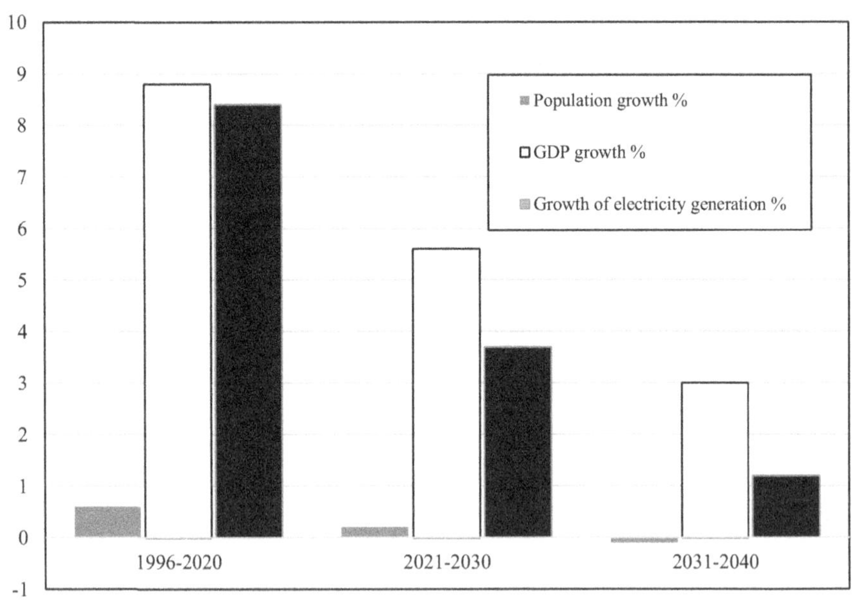

Fig. 6.2 China's population, GDP and electricity generation growth, actual and expected (%, per annum). *Source* United Nations (2020) Authors' estimates

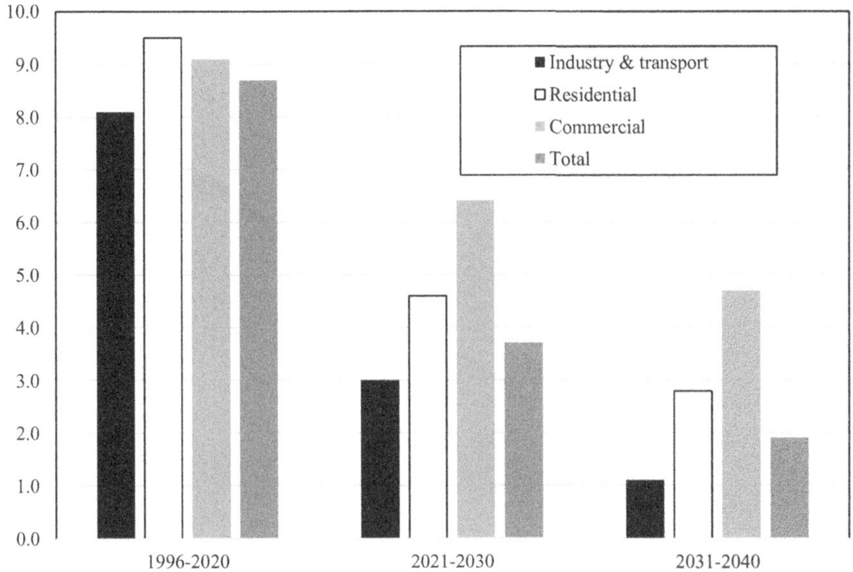

Fig. 6.3 Annual average growth of demand for electricity by sector, 1996–2040, actual and expected (% per annum). *Source* Authors' estimates

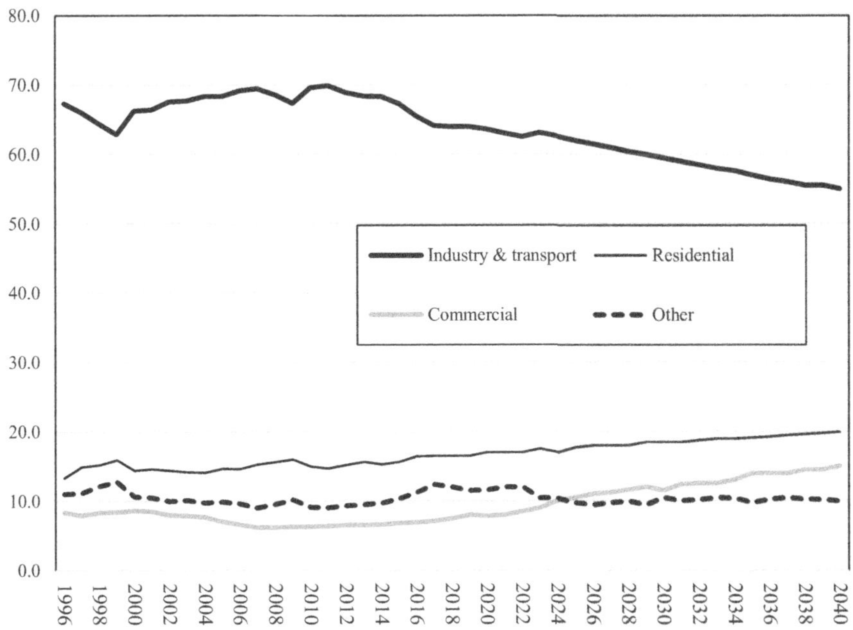

Fig. 6.4 Proportion of demand for electricity in China by sector, 1996–2040, actual and expected (%). *Source* Authors' estimates

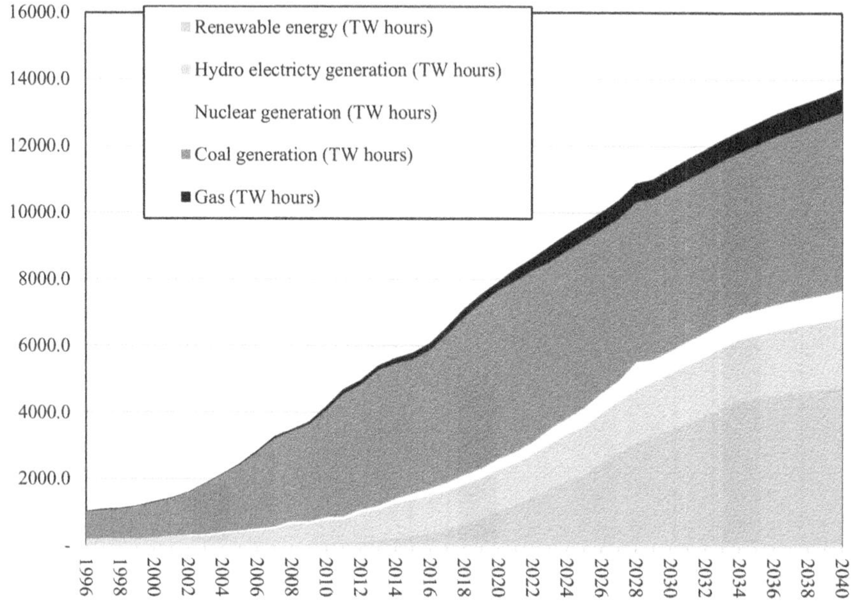

Fig. 6.5 Chinese production of electricity 1996–2040, actual and expected (TW hours). *Source* Authors' estimates

relative shares (Fig. 6.4) taken by each sector, but in the Chinese case, industry will still be by far the most important sector in 2040.

In terms of the source fuels for the generation of electricity, these are shown as actual (1996–2020) and forecast (2021–2040) in Figs. 6.5 and 6.6. As can be seen from Fig. 6.6, the proportion undertaken with coal-fired plant will decline from around 60% to around 40% by 2040, with the greatest growth in share taken up by non-hydro-renewable energy sources (wind and solar). A shift of this magnitude away from coal and towards renewable energy will have two main implications. The first is that it will mean that electricity prices will rise to pay for the high-cost renewable energy. This, in turn, will encourage a more efficient use of electricity and will temper demand growth somewhat.

The second implication is that the security of supply will become more unstable, as wind and solar are more intermittent in supply, given their dependence on wind and sun. To temper, the impact of both of these factors attention will need to be given to increasing the use of other non-coal sources of power such as natural gas, hydro and nuclear all of which can help to stabilise the supply of electricity. In addition, greater emphasis will need to be given to enhancing the interconnection between regions. Greater investment in transmission connections will help to create greater system stability as well as improve efficiency levels.

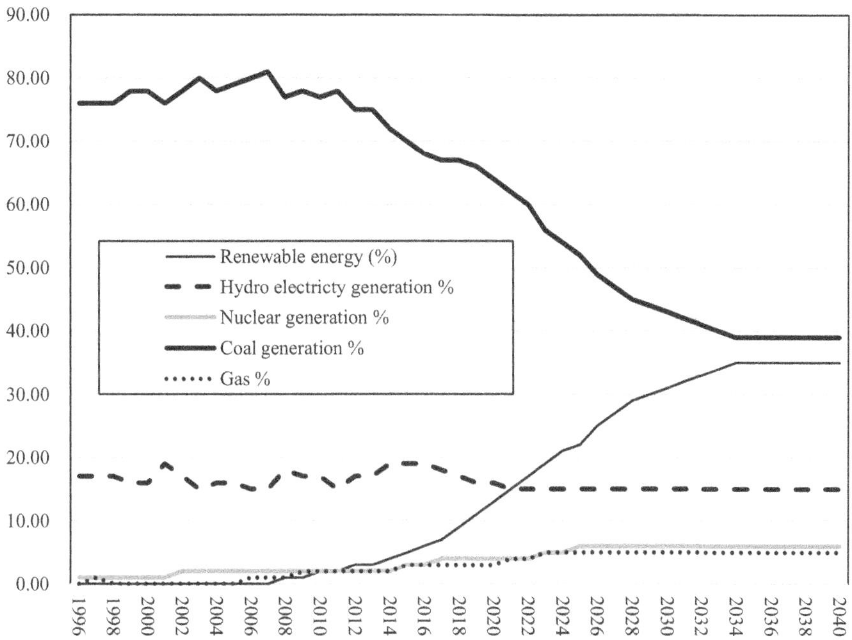

Fig. 6.6 Chinese production of electricity, by source 1996–2040, actual and expected (TW hours). *Source* Authors' estimates

6.6 Conclusion

The purpose of the chapter was to provide an outline of a model of electricity demand forecasting for the Chinese market. A range of variables is included in the model to capture the numerous factors that influence demand for electricity in the short and long run. Identifying these key variables is an important part of construction a model that can accurately forecast electricity demand growth. Once work on the model is complete, it is possible to use it to create greater clarity for investors who are engaged in investing in capital assets with a long life.

Although internationally there is a range of studies that look at forecasting electricity demand, in China this is less so, even though it is today the world's largest electricity industry. Any work, therefore, that contributes to the ongoing development of forecasting models for the Chinese electricity industry is of vital importance. What past work that has been undertaken in China so far has tended to focus on either the industrial or residential sectors of used aggregated national data.[9] By using all three major demand sectors, it is possible to progressively test the results of the model as it is developed and improve on its accuracy.

Finally, as demand growth in China has exhibited some volatility and as the Chinese Government is committed to diversifying energy sources in the sector the

[9]See, for instance, Holtedahl and Joutz (2004), Liao et al. (2017).

use of forecasting models will become more important in determining both policy and investment decisions.

Electricity consumption prediction is of great importance for developing and expanding a power system, fostering economic development and supporting people's daily activities. However, due to increasing demand in many areas and higher complexity, an accurate prediction for electricity consumption becomes more and more difficult. Therefore, designing an appropriate methodology for short-term prediction with limited data makes a great deal of sense.

The total and industrial electricity consumption is important indicators, not only reflecting economic development and industrial productivity, but also for developing energy strategies and associated environmental protection policies. Forecasted results indicate that China's total and industrial electricity consumption will continue to have a strong increasing trend in the upcoming years, but over time power energy sources and demand sources will change, bringing new challenges to the sector.

References

ABARE/Asia-Pacific Economic Cooperation Energy Working Group. (2002). *Deregulating energy markets in APEC: Economic and sectoral impacts.* Canberra ACT: APEC Secretariat.

Adams, F. G., & Shachmurove, Y. (2008). Modeling and forecasting energy consumption in China: Implications for Chinese energy demand and imports in 2020. *Energy Economics, 30,* 1263–1278.

Australia, Bureau of Resources and Energy Economics. (2014). *Australian energy projections to 2049–50.* Canberra, ACT: BREE.

Australian Gas Association/ABARE (1996). *Price elasticities of Australian energy demand.* AGA Research Paper No. 3, Canberra ACT: AGA.

Bhattachayya, S. (2011). *Energy economics: Concepts, issues and governance.* London: Springer.

Blackman, A., & Wu, X. (1999). Foreign direct investment in China's power sector: Trends, benefits and barriers. *Energy Policy, 27,* 695–711.

BP. (2019). *BP statistical review of worked energy.* London: BP.

Chow, G. C. (1993). Capital formation and economic growth in China. *Quarterly Journal of Economics, 108*(3), 809–842.

Cox, A. (1987). *Forecasting the substitutable components of energy demand in Australia.* ABARE Paper Presented at the Conference of Economists, Surfers Paradise, Queensland, 23–27 August.

Dergiades, T., & Tsoulfidis, L. (2008). Estimating residential demand for electricity in the United States, 1965–2006. *Energy Economics, 30,* 2722–2730.

Gianna, B. (2011). *Investing in energy: A primer on the economics of the energy industry.* Princeton: Wiley.

Holtedahl, P., & Joutz, F. L. (2004). Residential electricity demand in Taiwan. *Energy Economics, 26,* 201–224.

Liao, H., Cai, J. W., Yang, D. W., & Wei, Y. M. (2016). Why did the historical energy forecasting succeed or fail? A case study on IEA's projection. *Technological Forecasting Social Change, 107,* 90–6.

Liao, H., Liu, Y., Go, Y., Hao, Y., Ma, X. M., & Wang, K. (2017). Forecasting residential electricity demand in provincial China. *Environmental Science Pollution Research, 24,* 6414–6425.

Sadeghi Keyno, H. R., Ghaderi, F., Azade, A., & Razmi, J. (2009). Forecasting electricity consumption by clustering data in order to decline the periodic variable's affects and simplification the pattern. *Energy Conversation Management, 50,* 829–836.

Son, H., & Kim, C. (2016). Short-term forecasting of electricity demand for the residential sector using weather and social variables. *Resource Conservation Recycling, 123,* 200–207.

Steenhof, P. A., & Fulton, W. (2007). Factors affecting electricity generation in China: Current situation and prospects. *Technological Forecasting Social Change, 74,* 663–681.

United Nations. (2020). *World situation and prospects 2020.* New York: United Nations.

von Hirschhausen, C., & Andres, M. (2000). Long-term electricity demand in China—From quantitative to qualitative growth? *Energy Policy, 28,* 231–241.

Chapter 7
Conclusion: The Way Forward

7.1 Conclusion

The general purpose of this book has been to provide an examination of the present state of the electricity industry in China. In does so, looking back at its historical development, its current state and its future potential. The relationship between this industry and the broader energy sector in China was examined including the use of other sources of energy and its development.

It seems fairly obvious that the future of the electricity industry in China will be different, from the way it operated in the past, although heavily influenced by much of the developments of the past. For many years to come generation will be dominated by coal-fired plant, although it is expected that gradually natural gas, hydro, nuclear and renewables will become more important. The historical legacy of past ones is a heavy one in this industry, given the large-scale investments in the past in capital intensive plant. The links between the industry and the large demands of heavy industry for electrical power will also make help to make change a protracted one. In the short term growth in demand will be steady, although not as great as in the past, so the construction of new coal-fired plant is expected for some years at least. This will eventually change.

As China's population growth slows, so too will its economic growth and subsequently its growth in demand for electricity. Whereas, in the past, the key imperative has been to construct as much generation capacity as possible in order to meet strong demand growth, the future imperatives will increasingly be to achieve the greatest levels of efficiency possible, given slower growth in demand. This means that there will be a growing emphasis on more fully utilising existing generation capacity, which, in turn, will mean a growing interest in building transmission links between the regions. In addition, a growing interest in achieving greater reductions in carbon emissions will mean some replacement of the older, coal-fired plant with hydro and renewables.

M. Xiaoying and M. Abbott, *China's Electricity Industry*,
SpringerBriefs in Energy, https://doi.org/10.1007/978-3-030-53959-7_7

The other legacy of the past besides a reliance on coal-fired plant is the dominance of the industry by large-scale, government-owned companies. Although some separation has occurred, some deregulation of pricing and some encouragement given to new private investors in plant, the dominance of the state of much of the industry is a deterrence to new investors in non-coal-fired plant. Local provincial authorities have also shown a tendency to favour the interests of incumbents. If a greater diversity of fuel sources is to be achieved, along with lower carbon emissions is to be achieved further reform of the markets need to occur. At the same time, greater integration of the markets through the construction of transmission interconnections between regions will also be necessary.

The manufacturer's authorised representative in the EU is Springer Nature Customer Service Centre GmbH, Europaplatz 3, 69115 Heidelberg, Germany. If you have any concerns regarding our products, please contact ProductSafety@springernature.com

Printed and bound by CPI Group (UK) Ltd, Croydon, CR0 4YY

30/09/2026

02225575-0005